The Portable
Router Book

OTHER BOOKS BY THE AUTHOR

The Portable Router Book

2nd Edition

R.J. De Cristoforo

TAB Books
Division of McGraw-Hill, Inc.
Blue Ridge Summit, PA 17294-0850

SECOND EDITION
FIRST PRINTING

© 1994 by **TAB Books**.
TAB Books is a division of McGraw-Hill, Inc.

Library of Congress Cataloging-in-Publication Data

DeCristoforo, R.J.
    The portable router book / by R.J. De Cristoforo. — 2nd ed.
      p.   cm.
    Includes index.
    ISBN 0-07-016336-7   ISBN 0-07-016337-5 (pbk.)
    1. Routers (Tools)   2. Woodwork.   I. Title.
TT203.5.D43 1993
684'.083—dc20   93-38575
                                    CIP

Acquisitions editor: April D. Nolan,
Editorial team: Annette M. Testa, Book Editor
                Joanne M. Slike, Executive Editor
                Stacey R. Spurlock, Indexer
Production team: Katherine G. Brown, Director
                Susan E. Hansford, Typesetting
                Rose McFarland, Layout
                Kelly S. Christman, Proofreading
Design team: Jaclyn J. Boone, Designer
             Brian Allison, Associate Designer
Cover design: Carol Stickles, Allentown, Pa.   TAB1
Cover photograph: Bender and Bender Photography, Waldo, Oh.   4432

Contents

Introduction
to the first edition

I've produced a number of magazine stories about the portable router and have included chapters on it in books dealing with power tools. Many times I have introduced the topic by saying something like, "You don't know what you are missing if you don't own a portable router" or "Even if you own a portable router, it's possible that you are not exploring its capabilities." I feel that expressing these same thoughts is a good way to get started with this project.

You can't do justice to this interesting tool in a three- or four-page magazine article or even in a lengthy book chapter. That fact is what prompted the production of this book, which is dedicated exclusively to portable router techniques. For information concerning router projects, see *24 Router Projects—2nd edition*, (TAB #4488), by Percy W. Blandford.

The portable router is something of an enigma. Its mechanics are deceptively simple: a motor driving a shaft or spindle that has a chuck-type device at its free end. On its own, it can't do more than spin its chuck—a comparison you can make with a table saw or radial arm saw that isn't equipped with a saw blade or a drill press that doesn't have a boring bit. However, by adding cutters (router bits) or a host of accessories and jigs that you can make or buy to the router, the tool can be transformed into a multipurpose unit that allows anyone to accomplish a full range of practical woodworking chores and to do them professionally.

Too often, like the fairly common view of a stationary shaper, the tool is viewed as a means of producing decorative edges on projects such as tabletops, trays, picture frames, and so on. It can do such work superbly, but the limitation is like using a scroll saw just to produce puzzles.

The modern router has types, sizes, and price ranges to fit any level of interest and size of pocketbook. It just about eliminates the dividing line between amateur and expert in many areas. For example, with a dovetail jig, anyone can produce precisely fitted, classic dovetail joints. Some of the jigs limit the user to equally spaced units, but others allow spacing that suits the worker's view of how the connection should look. Thus, individual craftsmanship also plays a part. The work isn't accomplished simply by saying *abracadabra*, but the possibility of human error is greatly reduced and apprenticeship time is practically nil. If you follow the instructions supplied with the jigs and handle the router correctly, the first joint you make will be perfect. The same happy thought applies to other classic joints like the mortise-tenon and to more prosaic connections like dadoes, grooves, and rabbets.

The router is a super tool for forming woodworking joints, but there is much more. Its capabilities include shaping edges on project components either before or after assembly, hollowing out trays and chair seats, piercing for decorative effects, producing duplicate parts, leveling slabs, creating decorative panel routing, carving 3-D or bas-relief projects, routing patterns in your work, and so on.

Manufacturers have become aware of router potential and are now offering unique accessories that assist the woodworker with, among other things, making bowls, creating decorative work on spindle turnings, shaping controlled letters and numbers, duplicating stencils or drawings using pantographic devices, and even cutting precise screw threads in wood.

The router is a portable tool, so most times it is applied to work that is being held stationary in some fashion. There are times, however, when it is more convenient to use the tool like a stationary unit. The method of use is then reversed—that is, the work is applied to the tool. There are even situations, usually created by the types of router bits being used, where such an arrangement is mandatory. Cutters like Freud's "Panel Door Set," Zac Products' "The Door Shop," and Sears' "Crown Molding Kit" are never used in the freehand manner that is normally associated with portable router techniques.

Using the router in more than its hand-held capacity is no problem for anyone who wishes to fully utilize the tool. Many manufacturers offer a special stand in which the router can be installed and used like a stationary shaper. Chapter 12 shows a few commercial router/shaper stands and also offers construction drawings for a sophisticated, floor-model unit that you can make in your own shop.

The use of jigs, fixtures, and guides that you can make are as important to efficient and sometimes safer power tool use as the tools themselves. Often the homemade devices allow the tool to function in a manner that even the manufacturer didn't anticipate. Other times the jigs, fixtures, and guides serve to reduce, if not eliminate, the possibility of human error. This is an important factor, that is why guides should be used if at all possible. The user should adopt a mechanical means of guiding the tool to assure accuracy whenever possible. *The Portable Router Book* gives plans for many accuracy accessories that are useful and relatively simple to make. When you encounter a situation that calls for a special setup, give some thought to what can be done to provide automatic precision. This is especially important when the router is used to make duplicate pieces and when the same cut is required on many project components.

The portable router is considered to be a fairly safe tool, but in my mind, no tool, whether hand or power driven, is *safe*. It's important to remember that safety is as much in the mind as it is in the tool or, in particular, in the procedures you use with the tool. Anything that can cut wood and even harder materials can also cut you. Never lose your respect for the tool. The complacency that can accompany increased knowledge of the tool can spoil workshop fun. As many professionals as amateurs are hurt in workshops; expertise doesn't guarantee safety. Pay particular attention to chapter 2, obey the rules found in that chapter, and remember that "thinking" is your job. Even tools that can make some decisions for you electronically can't tell whether they are confronting wood or a finger.

Two adages of mine ...

"Think twice before cutting."

"Measure twice, cut once."

Introduction

The portable router concept continues to excite and inspire users and manufacturers. Since the publication of *The Portable Router Book* in 1987, the tool, while still "deceptively simple," has developed further to provide additional features that help woodworkers operate more conveniently and professionally. For example, plunge routers, that were rare in this country not too long ago, have become common. Similarly many routers now offer a fixed set of speeds or variable speeds, a factor that contributes to work-quality and safety. There are even heavy-duty units with impressive horse-power and new bits and accessories to complement your tools.

Improvements and advances of a tool ultimately prompt a revision of an "old" book, but that's not the only reason for completing this second edition. Continued use of a tool leads to ideas for improving homemade jigs and to conceiving new concepts for jigs. Some of the new jig concepts offered in this book include using the router as an overarm pin router, a motorized lathe chisel, a panel routing jig, an accurate depth gauge, center routing guides, and as a dowel tenoning jig.

Among the new commercial accessories that are covered in this second edition are those that allow the router to form *biscuit* joints, to create special moldings, and to form rosettes.

Also new in this revision are some joint forming accessories like the Keller Dovetail Templates, which are notable for their ease of use, and the Jointmaster, a group of integrated templates used for the production of connections like the mortise-tenon and finger-lap joints, and dovetail joints.

There are other noteworthy products highlighted in this second edition like the Trimtramp, an unusual name for a practical accessory that was originally designed for use with a portable power saw but is now suitable for a router, thanks to a special adapter kit.

So the portable router saga continues. No wonder it's one of the most popular woodworking power tools available today.

1

The tool

It's fairly certain that the first "router" was a hand tool—a hand plane or, specifically, a *hand router plane*. A more sophisticated version of the original tool is seen in the modern Stanley version in FIG. 1-1. The tool, designed for a two-hand grip, has a husky base and a means of securing a cutter for a specific depth of cut. The idea to create such a tool probably originated with someone who was using hand chisels to remove material to hollow a tray or to form a dado or groove. Hand planes were probably visualized the same way as the modern day router—a tool used for doing a chore with less effort and with mechanical control for better results.

If you substitute an electrically powered rotary cutter for the stationary blade of the hand tool, you'll discover the advantages of the basic portable router. That is, the depth of cut is still controlled and the router does the work while you concentrate on being creative.

ROUTER BACKGROUND

The invention of the electric router is credited to R. L. Carter, a respected wood and metal patternmaker with an inventive mind. Like many ingenious ideas, the powered router came about because of a chore that ordinarily required many tedious hours of work with a tool that resembled a spokeshave. The solution, which sounds simple, was to redesign a worm gear (removed from a barber's clipper) as a cutter and attach it directly to the shaft of an electric motor.

The improvisation worked impressively, especially after Carter added guides to provide accurate cut control. The new *hand shaper*, having solved the problem at hand, was then stored away for more than a year. Nephew Julius A. Yates rediscovered it to demonstrate to a cabinetmaker how much

1

1-1 Earlier versions of this hand router plane preceded the development of the modern electric router. Note that the tool's base has been extended with an auxiliary base. This "extra span" technique is also used with powered routers.

easier it was to use the electric tool in place of hand tools to carve some ornate curves in the walnut back of a sofa that was being constructed. The cabinetmaker became an enthusiastic first customer of the original electric router.

It was sometime later that the two men, working out of a garage machine shop, began to mass-produce the tool at about 15 a week. Even this large quantity of newly produced routers did not satisfy the avid, waiting customers. Factory space was increased, employees were added to the assembly line, and in 10 years, more than 100,000 tools were in use.

Stanley acquired the Carter business in 1929 and their improvements, plus those the originator had continued to develop over the years, resulted in the amazingly versatile machine that exists today. Stanley manufactured routers until about 1980 when their power tool division was passed on to the Bosch Power Tool Corporation.

Currently routers are being manufactured by such companies as Porter-Cable, Black & Decker, Sears, Roebuck and Co. under the Craftsman name, Milwaukee, Ryobi, Makita, Skil, and others. Rockwell, which was commonly seen on portable router nameplates after the company had acquired Porter-Cable, is no longer seen on the tools since that line of portable routers, together with other portable electric tools, has been returned to Porter-Cable.

There are a plethora of types and sizes of routers around. With such a wide variety of routers, anyone can acquire exactly the tool or tools needed to suit the job at hand. Many of the manufacturers offer exciting accessories, some of which are listed in TABLE 1-1. Acquiring router accessories enables you to utilize the tool beyond its basic functions.

Table 1-1 Portable router accessories.

Accessory	Description—Application
Bits	A host of styles for decorative or joinery work; available in high-speed steel, solid tungsten carbide, or carbide tipped; some Teflon coated; most have integral or ball-bearing pilots
Dovetail jigs and templates	Allow quick and accurate forming of various styles of dovetail joints; some units allow operator to decide joint spacing
Router/shaper tables	Router can be mounted and used like a stationary shaper; usually equipped with adjustable fences and guards; can be used to shape straight or curved edges
Hinge mortising	Adjustable templates for accurate routing of mortises for door hinges; commercial units popular with home builders; homemade jigs can be substituted
Planer attachment	Allows the router to be used like a portable, powered plane; not available for all routers
Edge guide	Guides the router for cuts parallel to an edge; supplied with the tool or is available as an extra-cost accessory
Circle cutting	Jig guides the router through circular grooves or disc cutting; available commercially, but easy to make
Template guides	Used in router's base; essential for many applications, including pattern routing, dovetail cutting, and template-guided cutting
Laminate trimming	Special bits for trimming laminates and similar material flush to edges; various types; usually solid carbide or carbide tipped
Bit sharpener	Available as router attachment; mostly for sharpening all-steel bits
Letter/number guides	Templates for routing house numbers and name plates; available in various sizes and styles
Pantograph	Guides router through duplication of designs, letters, numbers, and so on; most have enlarging or reducing capability
Panel jigs	Adjustable guides and templates for decorative grooving of doors, panels, and drawer fronts; commercial or homemade
Router Recreator*	Guides router through duplication of 3-D objects as well as letters and numbers
Router Crafter*	Lathe-type accessory for forming decorative furniture legs, posts, and similar components
Wood Threader**	Used for forming screw threads in any species of hard or soft wood; matching taps available
Bis-kit System*	Used with router for biscuit-type joints
Rosette Maker*	Accessory used with Craftsman router/shaper table to produce decorative rosettes

*Craftsman (Sears, Roebuck and Co.) products

**Beall Tool Company

BASIC NOMENCLATURE

The portable router typically consists of two major components: the upper housing, which contains the motor and motor-driven spindle, and a base, which receives the housing (FIG. 1-2). The base maintains the motor in vertical position and allows the motor to be raised or lowered (means differ from tool to tool), creating the cuffer-projection below the base or the depth of cut.

1-2 All portable routers consist of a motor and a base, but not all units come apart as pictured. When they do, parts can be purchased separately. Many users have an extra motor for use in a permanent position in a commercial or homemade jig.

The router's base assembly includes a removable subbase that is usually made of an opaque black plastic (FIG. 1-3). Most router users agree that while the "standard" subbase is sturdy enough for long-lived use, its opacity and an often minimum-size hole for the cutter to poke through make it difficult to see what the cutter is doing. This might not be a problem when the router is mechanically guided through a cut, but it is a nuisance when the router is guided freehand and the operator must follow a guide line. For this reason, manufacturers are now offering transparent bases (FIG. 1-4) as accessories.

A solution that is available to anyone is to make substitute subbases using a material like clear polycarbonate plastic. Acrylic-type plastics don't work too well because they scratch easily. The best material to use is Lexan. It is shatterproof and scratch-resistant and, with an occasional cleaning with an antistatic cleaning solution formulated especially for plastics, will stay clear and last as long as any conventional subbase.

To make transparent subbases, use the original subbase as a pattern size and a template for locating the attachment screw holes. Countersink or counterbore the holes so attachment screws do not project below the bottom surface of the homemade unit.

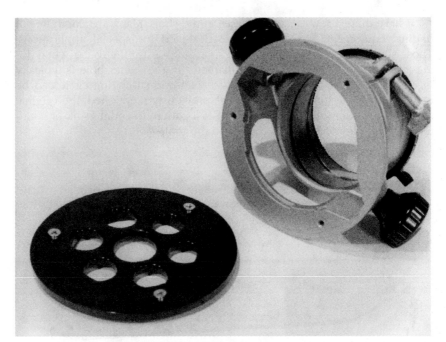

1-3 Routers have a removable subbase that is usually an opaque black plastic.

1-4 On many operations, the capacity of the subbase interferes with seeing what the cutter is doing. For this reason, manufacturers now offer a variety of transparent units of accessories.

I have quite a few Lexan subbases in the shop. Many are duplicates of original equipment, while others are designed for special applications. Those shown in FIG. 1-5 provide extra span support for the router when doing jobs like hollowing a tray. The diameter of the center hole is typically increased so that bits that have a larger-than-normal cutting circle can be used. As with the transparent subbases previously mentioned, the original subbase for your router should be used as a pattern to countersink or counterbore holes for accurate attachment to the tool.

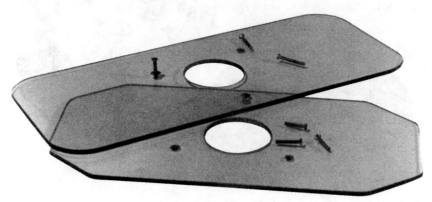

I-5 These extra-span subbases are made of Lexan, a shatterproof, scratch-resistant plastic that is available in many thicknesses. Such expanded subbases can be used for a variety of jobs.

THE GRIPPING END

Router bits are secured with a chuck, usually a split-collet type, that is attached at the free end of the motor spindle (FIG. 1-6). The collet compresses around the shank of the bit when a threaded locknut is tightened. Sometimes two wrenches are required: one to hold the spindle still while the other turns the locknut. Some routers are equipped with a spindle lock. This lock consists of a lever or push button that is used to hold the spindle in a fixed position thus allowing the user to need only a single wrench to turn the locknut.

In general, router sizes are typically specified by horsepower, but it's the size of the collet that determines the maximum shank size of cutters that can be used. For example, small routers with a ¼-inch collet are limited to cuffers with ¼-inch shanks, while larger units with ½-inch collets can handle larger bits. Often, a heavy-duty router can be equipped with collets of different size, say ¼ and ½ inch, or even ¼, ⅜, and ½ inch. Buying a heavy-duty router may be something to think about if your work scope covers a broad range of applications. Many production-type bits and special bits like those designed for use in a router/shaper setup have ½-inch shanks. Being able to handle heavy cutters as well as smaller shanked ones contributes much to a router's versatility.

A poor practice that can lead to injury is trying to increase depth of the cut by not inserting the shank of the cutter as far into the collet as it should

1-6 Bits are secured with a split collet that grips the shank of the cutter when a locknut is turned. Never try to use a cutter with a shank diameter that is smaller than the collet it is designed to take. Some routers can be used with various collet sizes.

go. The standard procedure is to insert the shank as far as it will go and then retract it ¹⁄₁₆ to ⅛ inch before securing the locknut. Depth of cut is always controlled by the height position of the motor housing in the base. Be sure the shank of the cutter, the collet, and the locknut are clean.

HANDLES

Full-size routers are designed for a two-hand grip, but there is a lot of variation in the size, shape, and even the placement of handles. A router's handle style is not a crucial factor when deciding upon a router; however, many expert woodworkers have preferences. It doesn't hurt to take a look at some typical designs that are available.

The Milwaukee unit in FIG. 1-7 has husky, open-top handles placed low on the tool's base. Although the shape of the handles might differ—some units will have knobs or closed-top handles—this is fairly typical. A possible disadvantage is that low-placed handles can cause some interference when the tool is guided along a straightedge or when the edge of the subbase must be moved along a special template. Solutions include using a thinner material as a guide so the handle can pass over it, rotating the router about 90 degrees so there is no interference, or, when possible, removing one of the handles. One of the favorable advantages for low-placed handles is that the operator can get wrists and forearms flat on the bench and have better control when doing freehand work. Figure 1-8 shows a Porter-Cable

1-7 One and a half horsepower Milwaukee router has sturdy, open-top handles. The 8½-pound unit has a no-load speed of 24,500 rpm and can be used with ¼-, ⅜-, and ½-inch collets. Note the position of the slide switch and the flat top for steady positioning when changing bits.

1-8 Porter-Cable 1½-horsepower tools have removable, knob-type handles located low on the base. This router turns at 22,000 rpm and can be used with three different size collets. Its net weight is 8¾ pounds.

tool with man-size control knobs positioned close to the base. With this router design, the knobs can be removed, providing the best of two worlds.

In some cases the placement of the handles is determined by the capacity in which the tool will function. The Black & Decker unit in FIG. 1-9 for example, is a plunge router so handles are affixed to the motor housing because they are used to control the vertical movement of the tool. The design of the Makita tool in FIG. 1-10 seems to oppose the view that low-placed handles provide better control. The handles are integral with the motor housing so, to some extent, their height above the work is affected by the depth-of-cut setting. Handles so placed with their spreading so wide beyond the motor can be a nuisance in tight areas and might be in the way when making homemade jigs and fixtures. An advantage of this particular unit is the placement of the on-off switch in one of the handles, a feature becoming

1-9 Black & Decker's 1½-horsepower unit has man-size, pass-through handles. This trim-looking tool weighs 8¼ pounds and has a speed of 25,000 rpm. Features include a spindle lock, built-in work light, handle trigger switch, and a removable chip deflector.

1-10 Makita's ¾-horsepower 23,000-rpm router has top-side handles that are integral with the motor housing. Contoured handles provide a comfortable grip, but location and spread can interfere on some operations. This 5-pound tool has a handle switch and can be used with ⅜- or ¼-inch collets.

common on many routers. This allows starting with both hands firmly on the tool and with the router placed where you wish to start the cut.

Top-side handles are not standard on Makita routers. The unit in FIG. 1-11 has a "D" handle plus a removable auxiliary knob. Good size "D" handles contribute a solid, secure feeling when the router is used, and they have the further advantage of usually containing a trigger on-off switch. A point against "D" handles made by some operators is that the design can interfere with adapting the tool for use with a router table. This might apply to ready-made router/shaper tables, but when you make your own (see chapter 12, The Router as a Shaper), the problem is solved by routing a groove to accommodate the handle.

There is more to router makeup than discussed so far. Figures 1-12 and 1-13, which are from Craftsman (Sears, Roebuck, and Co.) Owner's Manuals,

1-11 "D" handles, like the one on this Makita 1⅜-horsepower tool, are popular because they provide a solid, secure feeling. Another advantage is that they usually have a built-in trigger on-off switch. This 8-pound tool has a speed of 23,000 rpm and can work with ¼- and ½-inch cutter-shank diameters. The guide shown with the tool is an optional accessory.

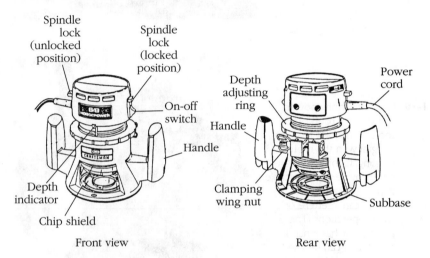

1-12 Basic nomenclature of a router. Routers are similar but have different operational features. Studying the Owner's Manual is essential for efficient use of the tool.

identify other router features. The point is, routers are the same, yet can be so different in the way depth-of-cut adjustments are made, switch locations, how cutters are secured, how the base is locked, and so on. How a router is used and the extent of its applications might be universal, but the instructions that are supplied with the router should be studied for that particular unit before use. By studying and restudying the Owner's Manual, you can appreciate and work hand in hand with the unit or units you own.

SIZE, POWER, SPEED

Routers differ not only in physical size and weight (FIG. 1-14), but also in horsepower and motor speed, more typically indicated as *revolutions per minute* (rpm). High speed, which can range from 15,000 rpm to better than 30,000 rpm, is traditional with portable routers and takes the credit for the smooth cuts you expect. Theoretically, the higher the speed, the smoother

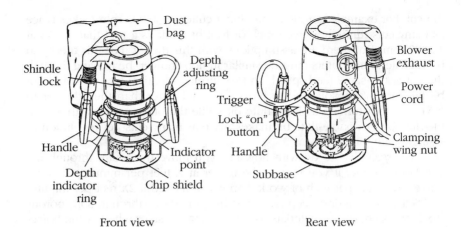

Front view   Rear view

1-13 Nomenclature of another router design. This one includes a dust collection system.

1-14 The "big" and "small" of it. The 3-horsepower unit on the left weighs almost 19 pounds. The more easily handled, ⅞-horsepower one on the right weighs 7 pounds. Each one, and others that fall between, play important roles in the portable router picture. The extent to which you plan to use the tool is the crucial factor when making a decision.

the cut. For example, a router spinning a cutter at 30,000 rpm makes twice as many cuts through a given inch or foot of work as a tool that works at 15,000 rpm. But there are assumptions with this theory. For this theory to be validated, the routers must be similar in horsepower, be moving through the cut at the same feed speed, and be turning identical bits. It would not be fair for comparison-sake to move one tool faster than the other or to have one tool turning a single-flute bit while the other works with a double-flute bit. A double-flute bit makes twice as many cuts per revolution as a single-flute bit.

Feed speed has to do with how fast you move the router through the cut. For example, if you moved the router at X feet per minute, it would make more cuts per inch of work than if you moved it 2X feet per minute.

When considering deep or heavy cuts, or passes through soft, porous stock or tough, dense materials, you must first consider whether the horsepower of the tool is adequate. Speed is a secondary factor to consider. For example, it's unfair to assume that the 1-horsepower unit shown in FIG. 1-15 could not possibly stand up to the operational "torture" that the 3-horsepower production router shown in FIG. 1-16 could abide. Often, the reverse is true. Some light-duty, low-horsepower routers have speeds of 30,000 rpm, while some bigger tools may only turn a cutter at 20,000 to 25,000 rpm. High speed does not always accompany impressive horsepower.

1-15 This 1-horsepower Makita router with ¼-inch collet capacity delivers 30,000 RPM. Features include a calibrated depth control ring, shaft lock so bit changing requires just one wrench, and light weight (5.5 pounds). The unit is typical of products with a motor that can be detached from the base.

The point: operationally, no router, or any power tool for that matter, should be used for work which is beyond its capacity. If you insist on making a router work beyond its capacity, the results will be poor and the motor will fail before its time. This doesn't mean that low-horsepower routers should be ignored. Much depends on the extent of your workshop interests. If your interests are more toward gardening or golf, and you do woodworking when you must or are occasionally inclined to, then a light-duty

1-16 This powerful Makita 3-horsepower plunge router has a 23,000 rpm speed and features an electric brake that stops the bit quickly when the trigger is released. It will accept ¼-inch or ½-inch collets, has a shaft lock, and a plunge capacity of 2⅜ inch. This unit weighs more than 12 pounds.

product will be a good choice. The small tool can, when necessary, even emulate larger ones by adapting special procedures. For example, while a husky router might be able to cut a ½-inch-wide × ½-inch-deep dado or groove in a single pass, a smaller unit can do the same by making repeat passes; that is, cutting to the same depth by going over the cut several times, an additional ⅛ inch or so deeper with each pass. Each cut allows the tool to work efficiently. You get there, it just takes more time.

SWITCH LOCATION

Most routers have either a trigger-type switch located in a handle (FIG. 1-17) or a toggle, rocker, or slide mechanism placed somewhere on the body of the motor (FIG. 1-18). There are pros and cons, sometimes arguments, about the best location. Major objections are voiced on the location of the motor-mounted switches, since the question of safety is concerned.

1-17 This Craftsman 2-horsepower router has a fairly typical, handle-located trigger switch that is lockable for continued use. Features include, speeds of 15,000 to 25,000, a dust collection system, a built-in work light, and a spindle lock. The tool weighs close to 10 pounds.

1-18 Toggle switches, like the one on this very popular, ⅞-horsepower, 22,000 rpm tool, are found on routers. The point against them has to do with safety. The important safety factor is knowing the characteristics of the tool and using it accordingly.

When the switch is located in a handle, the tool can be placed in firm operating position before it is activated. It can also be turned off and on at will during the procedure without relaxing your grip on the machine. Also, it's easier to oppose the motor's initial starting torque, which is generally characteristic of routers, when both hands are in holding position.

A disadvantage, unless the unit is equipped with two switches, is that it is difficult, if at all possible, to use the motor independently of the base. Separating the components is necessary when the motor alone is used somewhat like a high-speed grinder and when it is adapted for use with homemade jigs and some commercial accessories.

The switch arrangement that probably gets the most criticism is one that can't be used unless the operator removes one hand from the tool. This means that the starting torque must be opposed and the tool's work position held firm with one hand. Another concern with a unit with such a switch arrangement is that the switch could inadvertently be in the "on" position when it is plugged in. To combat this concern, follow the basic safety rule that applies to any power tool: check the on-off switch before connecting it to a power source. The word "inadvertent" in this case simply means forgetting that power tools can cut you as well as wood.

To help alleviate the concerns with the motor-mounted switch, Milwaukee designed a unit (see FIG. 1-8) where the switch is placed where it can be reached without relaxing the basic grip. The thumb of the hand gripping the right-side handle can reach up to move the slide switch.

PLUNGE ROUTERS

If you have ever used a conventional router to form a mortise or a stopped dado or groove, operations where the cutter must form its own starting hole, you'll quickly recognize the advantage of doing the job with a plunge router. For example, with a nonplunging router, a common technique for starting the aforementioned tasks is to hold the tool so its base is at an angle to the work surface and then slowly tilt it to get the bit started and until the base is in normal position. Human error and the safety factor of a projecting bit are both issues of concern.

Conversely, the plunge router moves vertically, usually on posts that are part of the base. Thus it can be firmly planted on the work and pressed down so that the bit enters the work at 90 degrees. The distance the bit projects can be preset and accurately maintained with a lever after full penetration. Because the units are spring loaded, the bit is easily retracted when the cut is complete by simply using the same lever.

Some of the tools have a turret-type device that allows presetting to several cut depths (FIG. 1-19). This is an asset when, for example, it's necessary to make a few passes to achieve a particular depth of cut or when a project requires dadoes of different depths.

Plunge routers are relatively new to us, and those that were initially introduced were more costly than conventional designs. But they are exciting tools and interest is growing. Even at this writing, there are choices in price, horsepower, and physical size (see FIGS. 1-20, 1-21, and 1-22).

A good conventional router *plus* a plunge router make a great team. If you are interested in purchasing only one router for your shop and wish to check out a plunging tool, consider other pertinent router factors. Can the motor be separated from the base? Can the tool be adapted for use in a router/shaper table? What are its capacities?

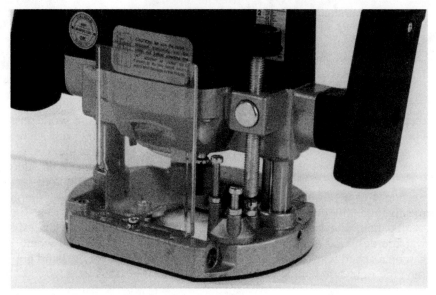

1-19 Most plunge routers have a rotating turret with adjustable screws so the tool can be preset for three specific plunge depths.

1-20 Skil's 1¾-horsepower plunge router has ¼-inch collet capacity and operates at 25,000 rpm. It incorporates a spindle lock and provides for wrench storage. The flat top design makes it easy to change bits.

ELECTRONICS

Routers haven't reached the point where they can function as independent robots, but many manufacturers are now thinking electronically and installing chips that help protect the router from abuse while guiding the user toward more efficient handling. The Black & Decker 1½-horsepower, plunge-type router provides a digital display with a depth-of-cut readout that can "talk" to you in either English or metric.

Being able to control a router's speed is an asset for both safety and performance. Many of the new products, like those in FIGS. 1-23 and 1-24, provide speed control. The control, usually a calibrated dial, provides either

1-21 The more powerful (2¼-horsepower) Skil plunge router works with ¼-inch or ½-inch collets and provides speeds of 10,000 to 23,000 rpm. It is equipped with a spindle lock and a fine adjustment depth stop. Note the wrench storage facility in the unit's top area.

1-22 Freud's new electronic plunge router ranks among the most powerful routers you can buy (3¼-horsepower). It is equipped with a ½-inch collet but includes a ¼-inch collet reducer. Variable speeds range from 8,000 to 22,000 rpm. Other features include a spindle lock and a "soft start."

1-23 The speed control dial on the Craftsman 2-horsepower router is located in the handle that houses the trigger switch. Speed range is from 15,000 to 25,000 rpm.

a set of specific speeds or an infinite number of speeds. When you consider the size of some of the router bits being offered, especially in the panel raising category, router speed-control is more a necessity than a convenience. There is more information on router speeds in chapter 2.

1-24 The Craftsman 3-horsepower plunge router has the speed dial located inside the right handle near the operator's hand. Speed range is from 10,000 to 25,000 rpm. Electronic speed control module senses the load applied to the motor and increases or decreases motor voltage to maintain desired rpm. The tool will accept ¼-, ⅜-, and ½-inch shank bits.

A good feature on some of the electronic routers is a slow, "soft start." The motor accelerates smoothly to its operating speed without the initial starting torque that's characteristic of conventional units.

In the future, you will probably see more electronic controls on all power tools. But for now, it's nice to know that the tool can do only what you wish it to.

TRIMMERS

Trimmers are designed specifically for working on plastic laminates, but there is nothing wrong with viewing them as palm-size routers. In fact, for some router work, like freehand sign making, detailed carving, and some operations where it is convenient to use the motor without the base, these small units are easier to control than full-size routers. All of the products are light in weight, in the area of 3 to 4 pounds, and operate at speeds as high as 30,000 rpm. Some are actually miniature routers, while others have features that make them particularly suitable for anyone, amateur or professional, who is doing a lot of work installing plastic laminates (FIG. 1-25). While motor design is similar, different base features make these units particularly suitable for special operations. For example, a tilting base is fine for trimming odd angled corners that are less or more than 90 degrees, while an offset drive spindle makes it possible to trim in or out of 90-degree corners.

The idea that trimmers can often be used like full-size routers is indicated by some manufacturers who offer router-type accessories as standard equipment or at extra cost (FIG. 1-26). The base of the tool is designed to accept an edge guide, which is used to guide the unit when making a cut parallel to an edge. The trimmer guide, mounted as shown in FIG. 1-27, is used for laminate trimming.

BASIC EQUIPMENT

I like to make a distinction between router accessories—items you can choose to do without—and basic equipment that is necessary for routine router operation. Basic equipment, in addition to router bits, includes edge

1-25 This array of 3.8-amp Porter-Cable trimmers ranges in weight from about 3 pounds to less than 5 pounds, works at speeds in the area of 28,000 rpm, and have a motor diameter of 3 inches. All of them have ¼-inch collet capacity. The one on the left is much like a miniature standard router, while the others have special operational features. The center one can be tilted. The third one has an offset drive spindle so it can be moved in or out of 90-degree corners.

1-26 Ryobi recognizes that a trimmer like its 3-pound 3.8-amp, 29,000-rpm unit, can often function like a full-size router. As standard equipment, they provide an edge guide that is used when trimming plastic laminates. Other accessories include template guides and high-speed steel or tungsten carbide cutters.

1-27 This is the way the trimming guide for the Ryobi router/trimmer is mounted. The cutting diameter of the bit and the outside diameter of the roller that rides against the edge of the work are the same. This is what produces the flush cutting for which trimmers are noted.

guides, trimming guides, and when applicable, additional collets. You're limited to a single bit shank diameter when the router is designed to work only with a ¼-inch collet. A larger tool, like the one in FIG. 1-28 that has a ½-inch capacity, is usually able to handle smaller collets as well. Sometimes basic equipment includes an extra collet, but it's often necessary to spend more money to acquire additional ones.

1-28 Heavy-duty routers, like this Skil unit, are usually equipped with ½-inch collets. Sometimes, smaller collets or sleeve adapters are supplied as basic parts. If not, they will be available as accessories.

The fact also applies to edge guides, examples of which are displayed in FIGS. 1-29 and 1-30. Special multipurpose guides, like the Black & Decker unit in FIG. 1-31, are always extra-cost items. With this unit, you can do more than just cut parallel to an edge. It can be used for decorating the edge of discs, forming circular grooves and arcs, and even some planing.

1-29 Edge guides, which direct the router through cuts that are parallel to an edge, may be supplied with the tool or offered as accessories. They are not interchangeable from tool to tool so check for compatibility with the tool you own before you buy.

1-30 Some edge guides can be fitted with a special roller guide. A major asset of the roller is that it will lead the router along curved edges.

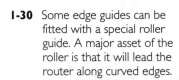

1-31 This multipurpose Black & Decker product is designed so the router can be secured to the disc.

THE DREMEL MOTO-TOOL

The Dremel Moto-Tool (FIG. 1-32), as the manufacturer states, is "A special tool for special jobs." Even the originator would be averse to classifying it as a "router." Any owner trying to use it for full-size router applications would be disappointed and would have a burned-out tool on his hands. Yet if the operator accepts its horsepower limitations and agrees that the proper way to use the tool is to combine its high speed with a light touch, allowing the cutter, not *forcing* it, to do the work, the Moto-Tool alone or with

1-32 The Dremel Moto-Tool is actually a high-speed grinder, but it can do router operations you wouldn't care to attempt with a full-size tool. It is available in single-speed models (about 30,000 rpm) or with variable speeds covering a range between 4,000 and 25,000 rpm.

accessories (like those shown in FIGS. 1-33 and 1-34) can function like its larger counterparts. For getting into tight places or shaping the details required on sculptures, the Moto-Tool, which handles almost like a pencil, has few equals.

To understand the Dremel product, compare it with an electric drill. The drill is a relatively low-speed but high-torque tool. In order to form a

1-33 The Moto-Tool can be equipped with various accessories that make it more suitable for router-type operations. Among them is this base which has a knob-adjusted, depth-of-cut control.

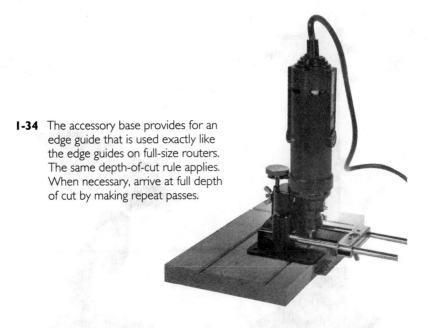

1-34 The accessory base provides for an edge guide that is used exactly like the edge guides on full-size routers. The same depth-of-cut rule applies. When necessary, arrive at full depth of cut by making repeat passes.

hole, the drill must be applied with some pressure. The Moto-Tool is a high-speed, low-torque tool. To use it efficiently, you utilize a high-speed/cutter combination. Applying excessive pressure would simply slow up the tool. Keeping the cutter in position and guiding it, allowing it to cut at its own pace, is the proper way to use this tool.

Currently, the ⅛-inch and ³⁄₃₂-inch collets for the Moto-Tool accommodate all cutting accessories (FIG. 1-35). These include high-speed burrs and tungsten carbide cutters, grinding points, steel saws, cutting wheels, wire brushes, polishing wheels, and more. There are also major accessories like those pictured in FIG. 1-36. Thus the tool can be used like a small drill press or held firmly at almost any angle so work can be applied to the cutting tool.

1-35 These are typical accessories that are available for the Moto-Tool. With them you can carve, polish, sand, clean, drill, do cutoff work, and more.

Of special interest to owners of "mini hand tools," like the Dremel and Fordham products, are the new "mini tool router bits" being offered by American Woodcraft Tools, Inc. under the Byrom name. These include straight, roundover, beading, cove, and rabbet bits, all of which have ⅛-inch shanks and cutting diameters that range from ⅛ inch to ³⁄₁₆ inch.

I-36 Accessories like this drill press and tool holder help make the Moto-Tool the versatile product it is. These units have been in my shop for a long time; the newer units have a modern look and additional features.

2

Safety

"It won't happen to me," is often the firm conviction of people who neglect to wear a seat belt when driving, who feel that a safety line when climbing a cliff is excess precaution, who point a knife toward themselves when cutting, and who generally feel that taking measures to avoid injury is only for the accident prone. There are also people who, because of initial respect and fear of a tool, start right, but become complacent with danger as expertise is acquired. Records maintained by safety organizations prove that as many, if not more, professionals are hurt using power tools as amateurs. The point is that practice and expertise with a tool that grows with experience does not make you immune to injury. Safety is as much in the mind as it is in the tool and in using correct procedures. Accepting that there is danger in a workshop is not different from being aware of the potential hazards of everyday life.

Tools are indifferent to what you present for them to cut. They can't think for you.

GENERAL SAFETY RULES

The safety equipment that is displayed in FIG. 2-1 should be standard equipment in all shops. It isn't difficult to convince workers about the logic of safety goggles, but the long-range effects of dust, and especially noise, are often overlooked.

Don't depend on prescription glasses for eye protection. Most safety goggles are made so they can be worn over lenses you normally wear. Don't store the goggles where it will be a nuisance to get them. Keep them exposed as a constant reminder.

The word "dust" leads many workers into thinking that dust masks are worn only when sanding, but many woodworking tools, including routers,

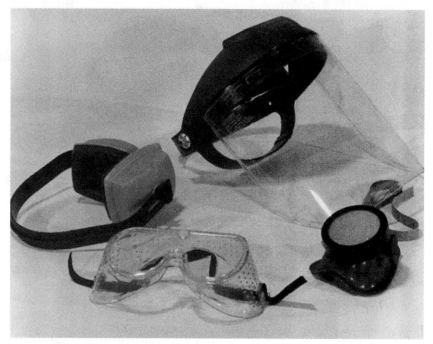

2-1 Shown here are safety goggles, a face mask, a dust mask, and headphone-type hearing protectors.

produce waste particles that should not enter your nose and lungs. Keep the filter in the mask clean and replace it as often as necessary.

The Occupational Safety and Health Administration (OSHA) sets safety standards for business. One of these standards is that a worker's exposure time to power equipment must be in relation to the level of noise frequency. The higher the sound frequency, the shorter the time a worker should be exposed to it. OSHA isn't looking over your shoulder in a home workshop, but the rule is important enough for all of us to obey voluntarily.

The damaging effects of high frequencies, especially those produced by routing and shaping operations, are cumulative. Every prolonged exposure affects hearing—perhaps only to a tiny degree—but it can build up to where hearing loss is obvious. Then it's too late to use a protective device. You don't want to block out sound completely. Good headphone-type hearing protectors, preferably of lightweight plastic, screen out hazardous frequencies while still allowing normal conversation. They also don't block out woodworking noises you should hear that might warn you of a malfunctioning tool or an incorrect operational procedure.

UNAUTHORIZED USE

Power tools are tempting items to children. Even some adults can't resist flicking a switch out of curiosity. Unplug the tool when you aren't using it. Store tools in cabinets, preferably ones that can be locked.

SHOPKEEPING AND SHOP DRESS

It's wise to have a special uniform for shop work. Trousers and shirts that fit snugly and nonslip shoes with steel toes make sense. Don't wear gloves, a necktie, or any article of clothing that can catch on an idle tool or one that is in use. You don't need jewelry in a workshop. Wristwatches, bracelets, rings, and any other adornments should be removed. Cover your hair, for safety and protection against dust, regardless of whether it is long or short.

A clean environment contributes to better work as well as safety. Maintain benches, tools, and accessories in pristine condition. Don't allow waste to accumulate on a workbench. Litter and wood scraps on the floor are dangerous slipping or tripping hazards. It's a good idea to have a wide shop broom and a shop-type vacuum cleaner for frequent use.

SHOP PRACTICE

Don't overreach, no matter what tool you are using or what the operation. It's very important to maintain firm footing and good balance at all times.

Don't work with dull cutting bits because they make it necessary to use more pressure to move the tool. This creates a situation where your hands might slip.

Disconnect the tool when it's necessary to change cutters or to add an accessory. Don't leave a power tool running when you turn to another chore regardless of how little time is involved. With a router, let the bit come to a stop before you set the tool down on its side or, if applicable, on its top.

Don't use tools as if they were stepladders. Serious injuries can result from slipping on a smooth surface or if the tool tips over. It's not likely that you will step on a portable router to stand taller, but I assume there will be other tools in the shop that you may stand on.

Recognize as a warning any operation where it becomes necessary to force a cutting action. Usually it indicates that the cutting tool is too dull or that you are trying to make a cut that is too deep for the tool to handle. Even super-horsepower routers have limits, so it's good practice to accomplish extra-deep or oversize cuts by making repeat passes.

Socializing and tool use are a bad combination. Visitors should not be welcome when you are using tools. It's also a good idea to educate friends and neighbors about the danger of barging into your shop when they hear a tool running. The sudden entry can startle you and cause an accident.

Staying alert and keeping your mind on the job are crucial safety precautions. Don't do shop work when you are tired, upset, or have had an alcoholic drink.

TOOL PRACTICE

Become familiar with the tool *before* using it, whether it's a replacement item or a new addition. Learn the tool's applications and, especially, its limitations. Don't use a tool for operations it was not designed to accomplish efficiently and safely. Most manufacturers warn against using accessories that were not specifically designed for the tools they produce. The admonition doesn't always hold, but it's a good idea to be cautious before being adventurous.

Never work on a piece of wood that is too small to be held or clamped securely. When, for example, you need a small piece with a shaped edge, do the shaping on a large workpiece and then cut off the part you need.

Be sure you know the procedure you must follow before you start an operation. By previewing the chore, you can anticipate possible problems, determine safest hand positions, and plan most efficient feed direction for the tool. Often, and especially with a new procedure, it's wise to go through a dry run; that is, go through the operation but with the tool turned off. Practices like this help make you tool-wise.

Keep the body of the tool and its handles dry, clean, and free from oil and grease. Clean tools with a lint-free cloth. Using solvents is not a good idea.

Some of the modern products have a built-in or, as in the case of the Skil router shown in FIG. 2-2, an add-on dust collection system. These usually require connection to a vacuum cleaner hose. When you organize your shop, take care that the added systems do not interfere with your freedom to move safely while working.

Check all parts of the tool, especially locking mechanisms, periodically to be sure no damage has occurred that could interfere with proper, safe operation. Switches that do not operate correctly should be replaced imme-

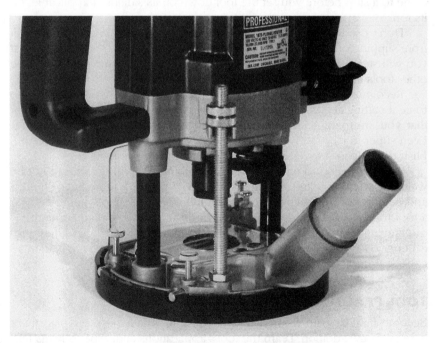

2-2 Some routers can be equipped with dust pick-up kit like this one.

diately, preferably by bringing the tool to an authorized service center. This suggestion is especially important if the tool is double insulated.

One of the most important safety factors I know is this—*always be a bit afraid of the tool.* Respect the fact that it is a machine. Whether you are in the apprenticeship stage or have advanced to a more knowledgeable plateau, don't ignore that operator responsibility is the main ingredient for a safe workshop.

ELECTRICAL CONSIDERATIONS

Many of the routers that are available today are double insulated. This is a safety concept in electric tools that makes it unnecessary for the tool to have a grounded three-wire power cord. The design of a double-insulated tool provides two complete sets of insulation to protect the user. That is, all exposed metal parts are isolated from the internal metal motor components with protecting insulation.

It is important with tools of this type that servicing be done with extreme care and only by qualified service technicians who have detailed knowledge of the system. An incorrect repair job can nullify the double insulation factor and place the user in danger.

Tools that are not double insulated will have a three-conductor cord and a three-prong grounding plug that should be used in a properly grounded outlet box. When the tool is designed for use on less than 150 volts, the plug resembles the one shown in FIG. 2-3. The grounding blade, which is longer than the two current carrying prongs, slips into the third, specially shaped hole in the outlet.

An adapter (FIG. 2-4) can be used to connect a three-prong grounding plug to a two-prong receptacle providing the outlet box is correctly grounded.

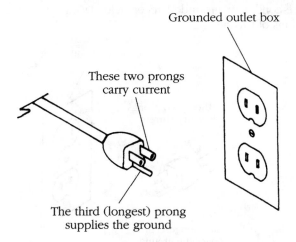

Grounded outlet box

These two prongs carry current

The third (longest) prong supplies the ground

2-3 Tools that are not double insulated and are designed for use on less than 150 volts have a plug with two parallel, current-carrying prongs plus a third one of different design, which is the grounding blade. Never cut off the grounding blade so you can use the plug in a two-hole receptacle.

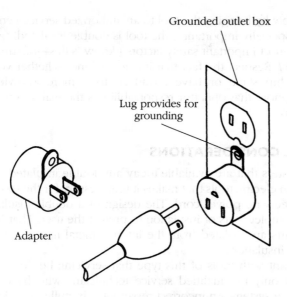

Grounded outlet box

Lug provides for
grounding

Adapter

2-4 It's possible to use a three-prong plug in a two-hole receptacle if you use this kind of adapter between the plug and the outlet. This system does not provide protection, however, unless the outlet box is properly grounded.

This adapter is not allowed in Canada. It is recommended that adapters be viewed as temporary measures for use only until a correctly grounded outlet box is installed, preferably by a qualified electrician.

If the tool is designed for use on 150 to 250 volts, the power cord has a plug that has two flat current-carrying prongs in tandem and a grounding blade that might be round or U-shaped. The plug works only with a proper mating, three-conductor grounded receptacle (FIG. 2-5). Adapters are *not* used with plugs of this design.

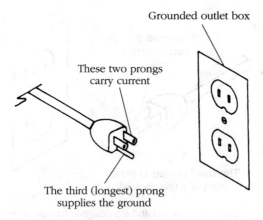

Grounded outlet box

These two prongs
carry current

The third (longest) prong
supplies the ground

2-5 Tools that are designed for use on 150 to 250 volts have this type of plug. It must be used with a proper mating, three-conductor grounded receptacle. No adapter is available for this plug design.

Never remove the grounding blade from a three-prong plug so you can use it in a two-prong outlet.

EXTENSION CORDS

The conductor size of an extension cord must be large enough to prevent an excessive voltage drop that will force the tool to work with less power and that can damage the motor. Recommended extension cord sizes in relation to length and the amperage or current rating of the tool are shown in Table 2-1. Extension cords that are suitable for outdoor use are marked with the suffix W-A, which follows the designation that tells the type of cord. SJTW-A is a typical example.

Table 2-1 Recommended extension cord sizes for portable electric tools.

Voltage		Length of cord (in feet)								
115 V		25	50	100	150	200	250	300	400	500
230 V		50	100	200	300	400	500	600	800	1000
Amperage rating (check nameplate)	0–2	18	18	18	16	16	14	14	12	12
	2–3	18	18	16	14	14	12	12	10	10
	3–4	18	18	16	14	12	12	10	10	8
	4–5	18	18	14	12	12	10	10	8	8
	5–6	18	16	14	12	10	10	8	8	6
	6–8	18	16	12	10	10	8	6	6	6
	8–10	18	14	12	10	8	8	6	6	4
	10–12	16	14	10	8	8	6	6	4	4
	12–14	16	12	10	8	6	6	6	4	2
	14–16	16	12	10	8	6	6	4	4	2
	16–18	14	12	8	8	6	4	4	2	2
	18–20	14	12	8	6	6	4	4	2	2

SAFETY BITS

The bit concept that is shown in FIG. 2-6 originated in Germany and is now being given a lot of consideration by American manufacturers, especially on oversize bits like panel raisers. The design limits the amount the bit cuts into the wood with each revolution. This lessens the tendency of the bit to dig in and grab the work. While the idea is not universal as yet, it does no harm to ask if the feature is present when you buy.

New vertical raised panel bits (FIG. 2-7), can fall into the "safety bit" category. Because of the vertical design and smaller cutting diameter (when compared to horizontal panel raisers), tip and surface feed rates are reduced considerably and horsepower requirements are lessened. The bits must be used in a router/shaper table and with a fence that is high enough

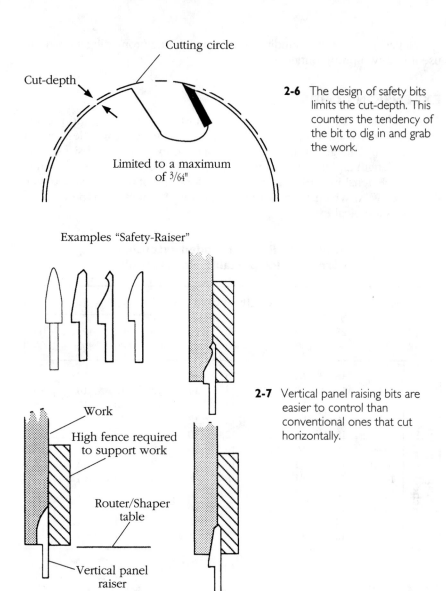

Cutting circle

Cut-depth

Limited to a maximum
of 3/64"

2-6 The design of safety bits
limits the cut-depth. This
counters the tendency of
the bit to dig in and grab
the work.

Examples "Safety-Raiser"

Work

High fence required
to support work

Router/Shaper
table

Vertical panel
raiser

2-7 Vertical panel raising bits are
easier to control than
conventional ones that cut
horizontally.

to support the panel. Profiles are "variable" simply by adjusting the projection of the bit above the table. Even with these more convenient bits, and especially when a low horsepower router is used, the full profile should be achieved by making repeat passes.

BIG BITS

The production of heavy-duty routers, as powerful as 3-horsepower and more, has motivated some manufacturers to offer bits that look more appropriate for an industrial stationary shaper than a portable router. Typical bits in this area are those designed for panel raising; some with as much as a 3½-inch-diameter cutting circle (FIG. 2-8). Just looking at them should warn anyone that they require a powerful router and one that allows adjustment of speed. My own feeling, regardless of the router in use, is that their use should be restricted to a router/shaper table and, even then, to rely on repeat passes to get to the full profile shape.

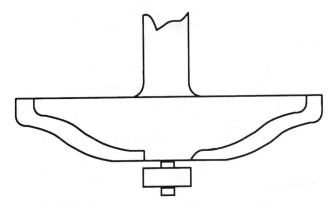

2-8 Conventional panel raising bits can have cutting diameters as great as 3½ inches. They must be used with heavy-duty routers and at low speeds. It's best to use them only in a router/shaper table.

SPEED CONTROL

There's no doubt that being able to control a routers speed adds to safety and work quality. Now, the economical variable speed control electronic devices, like the unit in FIG. 2-9, let you add the feature to any router that doesn't have built-in controls. To use the accessory, just plug it into an outlet and then plug the router into the unit. A 3-position rocker switch allows settings at "off," "variable speed," and "full speed." A special feature is an electronic feedback that maintains speed by increasing voltage to the motor as the load increases.

2-9 Accessory speed control turns any router into a variable speed tool. The unit is fused to protect it from overload and has a clip on the back so it can be hooked to a belt.

The general rule for router speeds is to *decrease* rpm as the diameter of the cutting circle *increases* (see Table 2-2).

Table 2-2
Suggested router speeds.

Diameter (cutting circle)	Max. speed (rpm)
1"	24,000
1¼" to 2"	18,000
2¼" to 2½"	16,000
3" to 3½"	12,000

FOOT SWITCH

Another economical accessory is a foot-powered switch that lets you turn the router on and off (FIG. 2-10). It's an especially useful item when you have to work with a router whose switch is not conveniently located in a handle, and when both your hands must be occupied with controlling the router and the work.

2-10 A foot switch that will turn the router on or off can add to safety since it allows you to use both hands to keep the router and work secure. Be sure that the switch capacity matches the amperage rating of the router.

SPECIAL CONSIDERATIONS

Safety factors that might apply especially to the portable router will be pointed out as you get into using the tool. The following is a brief summary of some of those safety factors:

- Always check to be sure the collet locknut is securely tightened before using the "on" switch. A loose bit can become a harmful projectile.
- Be sure the switch is in the "off" position before plugging into a power source.

- Don't have the tool plugged in when changing bits, making adjustments, or removing the base. Hold the tool firmly when you turn it off until the bit stops rotating. Then place it on its side with the bit pointing away from you. Don't make a bit change immediately after finishing a cut. The bit might be very hot.
- Check workpieces to be sure that nails or other foreign objects are not in the line of cut. Use carbide-tipped bits when working on nonwood materials, like plastics and plastic laminates, and on man-made products, like particleboard, hardboard, and plywood. High-speed steel bits work alright, but tungsten carbide blades cut smoother and stand up longer under abrasive abuse.
- Hold the router securely when you turn it on to oppose the twisting action (torque) that occurs. A two-handed grip is preferable. When this isn't possible, be sure to anticipate the initial action and be especially firm with one hand.
- Don't make depth-of-cut adjustments while the motor is running.
- Be sure the cord is free and can't snag on some nearby object during operation or trail across the cutting path.

3

Router bits

The portable router is a fascinating tool, but without bits it's like an automobile without wheels and can't do more than whir (FIG. 3-1). Router bits are the stars of the show and are as intriguing as the tool itself. The number of bits that are available can be bewildering, even intimidating, and others that contribute to the versatility of the router are constantly being added. Among the latest bits to appear are those that allow the router owner to produce all the shapes that are required to make professional paneled doors.

3-1 Router bits are essential to portable router use. The pictured bits are just a small example of what is available. Displayed are a few straight bits and two that have ball-bearing pilots.

These bits, having ½-inch-diameter shanks, are used with a heavy-duty router that is mounted in a shaper-type table.

Router bits can be placed in specific categories; for example, there are decorative bits and practical or utility bits. What you can accomplish with any bit, however, depends on whether you use it only for what it was designed to do or whether you wish to be more creative with the bit. Variations in the cut design made with a single bit are possible by following your first cut with a second one after the depth of cut or width of cut has been changed. The standard shape of one cutter takes on a new look when you repeat the operation using another cutter to add detail. This kind of work can't be done haphazardly. It's always best to draw on tracing paper the profiles of the cutters you plan to use so results can be previewed by overlaying the drawings.

With decorative bits, the cutters often produce classic molding forms (FIG. 3-2). Here too you can employ some control to alter results. For example, if the bit is guided by the pilot, you get a full width of cut, but can opt for a depth of cut that puts only part of the cutter's profile into play. The

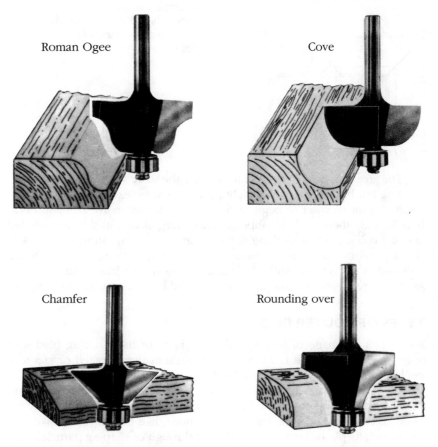

Roman Ogee

Cove

Chamfer

Rounding over

3-2 Many available bits, classified as decorative bits, are used to produce classic molding forms. When working with these bits, it is not imperative to always use the full profile of the cutter.

width of the cut can also be varied by ignoring the bit's pilot and moving the router through the cut by using an edge guide. Figure 3-3 shows examples of how shapes formed with decorative cutters and grooves formed with slotting cutters can be combined to fabricate framed panels.

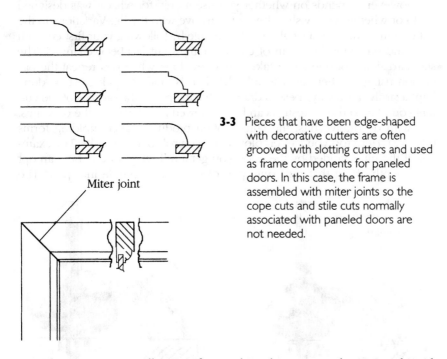

3-3 Pieces that have been edge-shaped with decorative cutters are often grooved with slotting cutters and used as frame components for paneled doors. In this case, the frame is assembled with miter joints so the cope cuts and stile cuts normally associated with paneled doors are not needed.

Miter joint

The greater your collection of router bits, the more work you can do with the tool, but buying everything right away would be exorbitant and unnecessary. It's much wiser to begin with straight and decorative bits and occasionally add others, such as rabbeting and dovetails bits, as they are needed until the collection is something like the fairly typical one shown in FIG. 3-4.

Beyond this assortment there are "special" bits, some of which are sketched in FIG. 3-5, and still others that will be shown later in this chapter. As always, what you buy should be determined by what you are working on.

TYPES OF ROUTER BITS

Router bits are made of high-speed steel (HSS) or have cutting blades of tungsten carbide, which is a tough, man-made material brazed onto a tool steel base (FIG. 3-6). Tungsten carbide bits can cost two to three times as much as their all-steel brothers, but they cut smoother and hold a cuffing edge 15 to 25 times longer. Paying more is often logical then, especially for "workhorse" bits. For some router applications, like trimming abrasive materials such as plastic laminate and cutting through or shaping particleboard or hardboard, tungsten carbide bits are almost mandatory. Better-quality

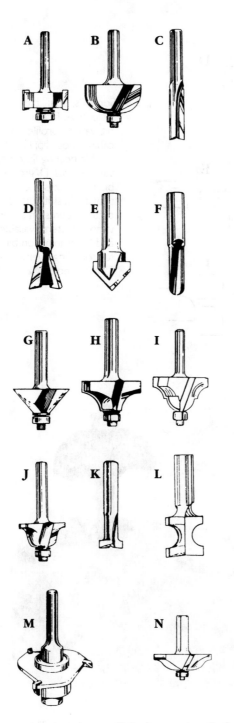

A. Rabbeting

B. Cove

C. Straight

D. Dovetail

E. V-groove

F. Veining

G. Chamfer

H. Beading

I. Cove and bead

J. Ogee

K. Key slot

L. Round edge

M. Slotting

N. Panel raiser

3-4 An assortment of router bits.

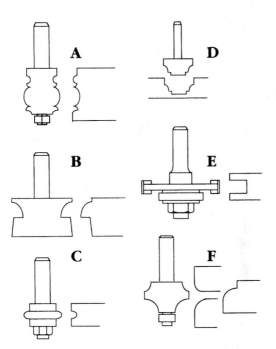

3-5 Some "special" bits include: A. face molding bit, various profiles; B. cabinet door lip; C. double bearing flute bit, various sizes; D. groove bit, many sizes and profiles; E. slot cutters, cutter mounts on arbor, various cutters available; and F. combination bit, for rounding over and forming moldings.

3-6 The bit on the left with the integral pilot is made of high-speed steel. The one on the right with a ball-bearing pilot has a tool steel body with brazed-on tungsten carbide blades.

carbide bits have edges that are thick enough to stand up to regrinding as many as 15 times, while cheaper ones won't take more than four or five. Also, quality units can be retipped, which often costs less than a replacement.

Some small bits, where it might be difficult to braze carbide cutting edges in place, are made of solid carbide (FIG. 3-7). There is a factor to be aware of regarding tungsten carbide. It is a very tough material, but it is also brittle. Cutting edges can be chipped or broken if the bits are not carefully stored or are banged against hard surfaces.

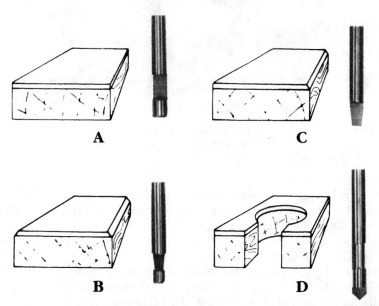

3-7 Some cutters that are made of solid carbide include: A. solid carbide flush trimmer with integral pilot; B. solid carbide trimmer with integral pilot, used to produce a beveled edge; C. multiuse trimmer used for square or beveled edges; and D. flush trimmer equipped with piercing point to start its own hole.

Carbide bits are not the only way to go. There is nothing wrong with buying good HSS bits, especially when selecting units you will only use occasionally.

There are other bits that home craftsmen won't use or can't use because the cost is too high to be justified. For example, industry is now being offered disposable bits that can be extremely low-cost units if purchased in quantity. When these bits become dull, they are simply discarded. Diamond bits are cuffers that have polycrystalline diamonds (man-made diamond crystals) bonded by special means onto a carbide base material. This design holds keen cutting edges up to 400 times longer than carbide, so it's cost effective for production work. However, would you care to spend $1,000 for, say, a ½-inch-diameter bit?

BIT DESIGNS

In addition to router bits being grouped in terms of the material they are made of, there are also categories like units that are one piece with integral pilots; units that consist of a shank, cutting blade, and screw-on pilot; and units equipped with removable ball-bearing pilots (FIG. 3-8).

3-8 Basic bit designs pictured include (from the right): one-piece HSS unit with integral pilot; assembly consisting of arbor, cutter and pilot; and two bits with tungsten carbide blades and ball-bearing pilots.

The *pilot* on a bit is what is used to guide it through a cut. The router is moved so the pilot bears against the edge of the workpiece throughout the pass. On one-piece bits, the pilot is turning just as fast as the cutting blades and creates considerable friction that often results in discolored work edges. When the operator applies too much pressure to keep the pilot in contact or when he pauses somewhere during the pass, the pilot can actually indent the work edge, especially on soft wood. Because the cutting blades follow the travel of the pilot, the occurrence causes a flaw. A dull cutter exaggerates the potential for errors because more feed pressure is required to keep the router moving. Good practice calls for sharp cutters, a steady feed speed, and only enough pressure that is needed to make the cut, resulting in burn-free edges and less effort on your part to move the

3-9 Regardless of the rpm of the cutter, ball-bearing pilots rotate in tune with feed speed, which is how fast you move the router to make a cut. Thus, the friction that is caused by an integral pilot, which turns just as fast as the cutter, is eliminated.

tool (FIG. 3-9). There is also the option, often overlooked, of using different diameter bearings. This contributes a degree of control over how wide a cut the bit will make. An example of using a different diameter bearing with a 1¼-inch rabbeting bit, is demonstrated in FIG. 3-10.

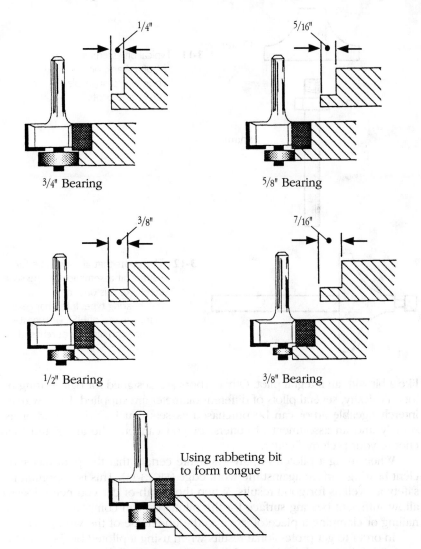

3-10 Various cuts that can be made with the same ¼-inch-diameter rabbeting bit by using bearings of different diameters.

Bit assemblies, which consist of the components shown in FIG. 3-11, look exactly like a one-piece bit when the parts are put together (FIG. 3-12). The advantage is that a single interchangeable arbor can be used with a variety of cutting blades. Blades are available in HSS or carbide-tipped designs. Some of the arbors are made to take solid steel pilots so the assembled unit works

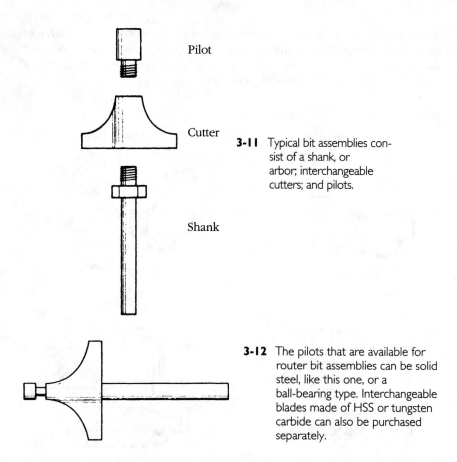

Pilot

Cutter

3-11 Typical bit assemblies consist of a shank, or arbor; interchangeable cutters; and pilots.

Shank

3-12 The pilots that are available for router bit assemblies can be solid steel, like this one, or a ball-bearing type. Interchangeable blades made of HSS or tungsten carbide can also be purchased separately.

like a bit with an integral pilot. Other arbors are designed for ball-bearing pilots. Typically, several pilots of different diameter are supplied. Units with an interchangeable arbor can be purchased as sets that include the arbor assembly and an assortment of cutters, or you can buy the arbor and then choose your preferred cutter.

When using a piloted bit, always make certain that the pilot has sufficient bearing surface against the work edge (FIG. 3-13). This is important for safety as well as for good results. When the depth of cut you need doesn't allow sufficient bearing surface for the pilot, you can compensate by tack-nailing or clamping a piece of stock to the underside of the workpiece.

In order to get professional results when using a piloted bit, be sure the work edge is smooth and free of flaws. The pilot will faithfully follow any roughness, bump, or crevice and guide the cutter to duplicate it. Also be certain that the pilot, integral or otherwise, is smooth and free of any wood residue.

PILOTLESS BITS

An assortment of pilotless bits is shown in FIG. 3-14. Many of these, and others that are not illustrated, can be classified as decorative bits. Some, like straight

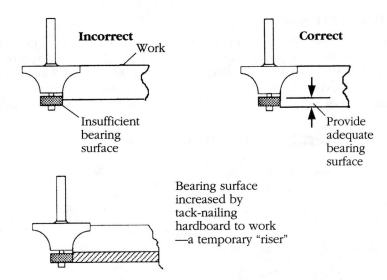

3-13 When using a piloted bit, it's important for quality work and safety for the pilot to have sufficient bearing surface against the work edge.

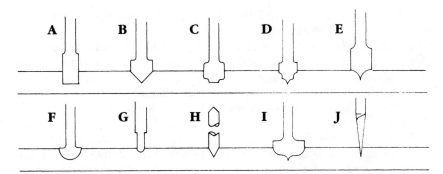

3-14 This assortment of pilotless bits includes: A. straight, B. V-groove, C. ogee, D. classic, E. quarter round, F. core box, G. veining, H. double-end vee, I. point cutting ogee, and J. tapered carving bit.

bits, can be used for fancy grooving, but they are essential for practical applications, such as shaping joinery forms like dadoes and rabbets as well. Regardless of the bit's general classification, it's what you are doing with a bit that defines it as a decorative or utility cutter. The V-groove bit, which is often used for simple or intersecting surface cuts, can be used to do some chamfering. Bits like the quarter round, ogee, and the core box bit (in the pilotless area) might be classified as surface-cutting bits, but they can be used to form a decorative detail on an edge as long as the router is properly guided.

Characteristic of the pilotless bit cutters is that the operation must be organized so the router itself is guided. This can be accomplished by many methods. The simplest and most often used is a commercial edge guide or a straight strip of wood clamped to the work against which the subbase of the router bears as the cut is made.

ROUTER BIT MAKEUP

Figure 3-15 shows the tip of a router bit and identifies its parts. The cutting circle is the actual working area of the bit. The shank diameter may be the same, but many times it is smaller. For example, a bit with a ½-inch cuffing circle can have a ¼-inch shank. When the shank diameter is greater than ¼-inch, you can assume that the bit should be used in a heavy-duty router.

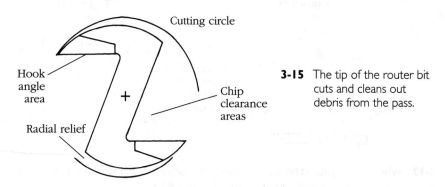

3-15 The tip of the router bit cuts and cleans out debris from the pass.

Shanks on quality bits are fully hardened and precision ground to a tolerance of 0.002. High hook angles plus adequate chip clearance areas (sometimes called *gullets*) provide for cleaner, cooler cutting. Radial relief provides clearance so only the cutting edges of the bits make constant contact with the work. Without it, the bit does as much rubbing as cutting. Bits with carbide inserts must be designed with a wall thickness that adequately backs up the brazed on blade. The best ball-bearing pilots are double shielded to protect the bearings from dust.

Generally, single-flute bits are stronger than double-flute bits (FIG. 3-16) but assuming some factors, like tool rpm and feed speed, they make half as many cuts per inch of work as a double-flute version. Most times the end of the bit is relieved so it can, when necessary, enter the work at a 90-degree angle (plunging).

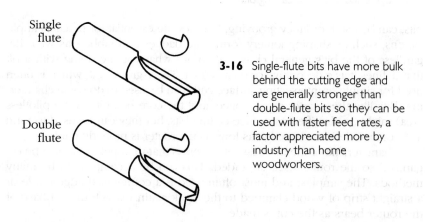

3-16 Single-flute bits have more bulk behind the cutting edge and are generally stronger than double-flute bits so they can be used with faster feed rates, a factor appreciated more by industry than home woodworkers.

Stagger tooth bits, like the carbide-tipped example in FIG. 3-17, are designed for balance and strength. They are used extensively by manufacturers of products whose components might be of composition materials like chipboard, particleboard, and plywood.

3-17 Stagger tooth bits with tungsten carbide blades stand up to the abrasive action that is present when working on materials like particleboard, hardboard, and plywood.

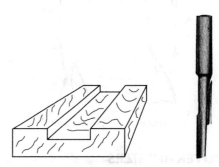

Other "straight" bits, some of which are just appearing and which are probably of greater interest to industry than the home craftsmen, include:

- Bits with concave flutes (O-flutes) that, because of additional strength, can be fed faster through solid wood.
- Chip-breaker bits that cut fast through abrasive, dense materials.
- Down-cutting or up-cutting spiral bits that push waste to below the work or up to its surface. Spiral bits cut constantly unlike, say, a single-flute bit that makes one cut per revolution (see FIGS. 3-18 and 3-19).
- Shear-cut bits look like straight flute bits but have inclined cutting edges. The manufacturer says that they are very free-cutting and require less horsepower for efficient results.

3-18 Spiral bits combine effective chip removal and smooth cuts.

3-19 Spiral bits are designed to move chips either up or down. The "up" spiral bits are especially suitable when cutting mortises or any cavity that does not pass through the work.

Up spiral   Down spiral

ROUTER BIT SIZES

When buying bits, you have a choice not only in configuration of the bit, but also in the size of the cut that the bit will produce. Other factors are total length, cutting length, and shank length.

Figures 3-20 and 3-21, which were copied from a Freud catalog, show typical offerings in the area of straight bits and dovetail bits. This also applies to decorative bits. For example, and again quoting from one manufacturer's catalog, the radius of cove bits can start at $\frac{3}{16}$ inch and go to $\frac{1}{2}$ inch, while the large diameter begins with ¾ inch and extends to 1⅜ inch. In other

Cutting diameter	Cutting length	Overall length	Shank length
1/8	3/8	13/4	11/4
3/16	1/2	2	11/4
1/4	2	11/4	
1/4	3/4	21/4	11/4
1/4	1	21/2	11/4
9/32	1	3	11/4
5/16	1	21/2	11/4
5/16	11/4	23/4	11/4
3/8	1	21/2	11/4
3/8	11/4	21/4	11/4
7/16	1	21/8	11/4
1/2	3/4	21/8	11/4
1/2	1	21/8	11/4
9/16	3/4	21/8	11/4
5/8	3/4	21/8	11/4
3/4	3/4	21/8	11/4
1	3/4	21/8	11/4

(Dimensions in inches)

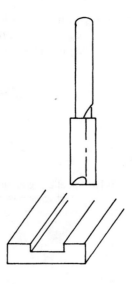

3-20 The number of router bits increases because of the various sizes that are available in a single bit design.

Slope degree	Large diameter	Depth of cut	Shank diameter	Overall length
9°	3/8	3/8	1/4	1 5/8
14°	1/2	1/2	1/4	1 5/8
9°	3/8	3/8	1/2	2
14°	1/2	1/2	1/2	2 1/8

(Dimensions in inches)

3-21 Various sizes are available in most bit designs, including dovetail cutters. Many times, similar bit characteristics are available in ¼- and ½-inch shank diameter units.

words, the smaller the radius of the bit, the smaller its large diameter. For example, a ³⁄₁₆-inch-radius bit will have a large diameter of ¾ inch, while a ½-inch radius bit has a large diameter of 1⅜ inch. Size considerations of other bits, like ogee, beading, and rounding over bits, can be described in similar fashion.

CMT Tools is now offering a wide variety of angles for their chamfer bits (FIG. 3-22). The bits (FIG. 3-23) are very precise cutters and can be used

3-22 Chamfer bits with various cutting angles are now available. They can be used to produce chamfers or full bevels.

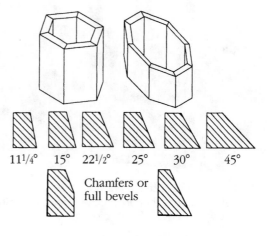

11 1/4° 15° 22 1/2° 25° 30° 45°

Chamfers or full bevels

3-23 Chamfer bits, like these from CMT Tools, are made to extremely tight tolerances. They must be in order to produce tight joints when used for multisided projects.

for simple chamfering or full bevels to produce multi-sided polygons. Anyone who has adjusted a table saw blade for the precise angle needed for a 6-sided or 12-sided planter box, for example, is aware of the frustrations. The chamfer bits are a happy alternative.

BUYING BITS IN SETS

There are advantages and disadvantages to buying bits in sets (FIG. 3-24). Getting a dozen or so bits at once enables you to do general router operations easily. The cost of a set is much less than the total price of the same bits pur-

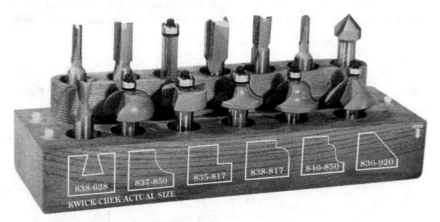

3-24 Buying bits in sets is practical and economical. These, from CMT Tools, are as handsome as they are precise. They are all carbide tipped and have a baked on, bright orange, TEFLON coating.

chased individually. Sets come in wood or plastic containers so the bits can be safely stored. A disadvantage, however, is that you may get more bits than your work scope justifies or designs you might ordinarily do without.

There are bits that must be purchased in sets because each bit does a job that is necessary for a satisfactory final result. The Sears Crown Molding Kit, shown in FIG. 3-25, is one example. Finding molding in a particular wood species is difficult if not impossible; therefore, with an assortment of bits like those in this set, you can custom-make moldings using wood from your wall paneling or furniture project. The kit contains bits for V-grooves, end coves, beading and chamfering, cove box, plus an adapter (arbor) and several pilots. Each bit is shaped to form part of a pattern so a sequence of passes, like those in FIG. 3-26, results in a profile that duplicates a standard molding or an exclusive one that you create (FIG. 3-27). The bits *must* be used in a router that is mounted in a router/shaper table. The shank diameters of the bits and adapter in the Sears Crown Molding Kit are ¼ inch.

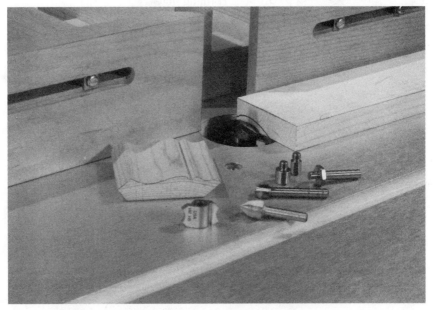

3-25 The Sears Crown Molding Kit includes all the cutters required to produce various molding designs. The bits must be used in a router/shaper.

Making cope cuts, forming stiles and rails, and producing a particular edge on panels are all cuts that are required in the construction of framed paneled doors and have always been associated with a stationary shaper (FIG. 3-28). Now several manufacturers are offering husky router bits so that anyone with a heavy-duty router can accomplish the same chores in a professional manner (FIG. 3-29). The only catch is that the tool must be used in a router/shaper table.

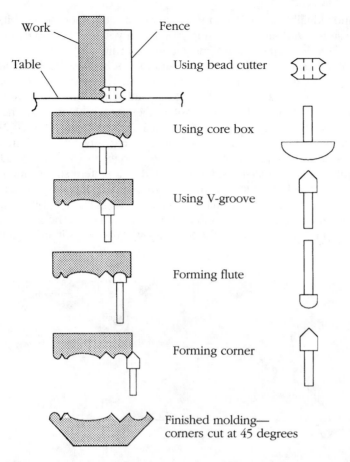

Work   Fence

Table   Using bead cutter

Using core box

Using V-groove

Forming flute

Forming corner

Finished molding—
corners cut at 45 degrees

3-26 With the Sears Crown Molding Kit, you can duplicate classic molding forms or create your own designs.

Zac Products, Inc., offers a set of bits called The Door Shop (FIG. 3-30). Figure 3-31 is a quick look at how these cutters are used. Note that some of the cutters do double-duty and that where the cut depends on the cutter's height above the table. This permits some variation and also allows working on stock of various thicknesses.

It's important to remember that bits in this area must not be treated lightly. The size of some of them alone should be enough to inspire respect and caution. More about paneled door making in chapter 12, The Router as a Shaper.

SPECIAL BITS

I use the word "special" casually for bits that might be needed infrequently and for those that are made for a specific chore. Anyone involved in a considerable amount of work with plastic laminates would not be without carbide-tipped plunge bits and laminate trimmers. Because they are necessary

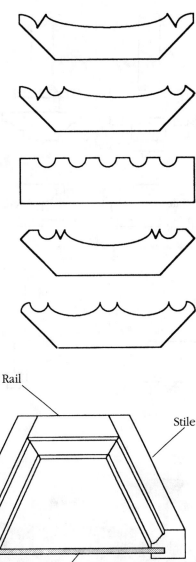

3-27 Examples of moldings that can be produced using the Sears Crown Molding Kit.

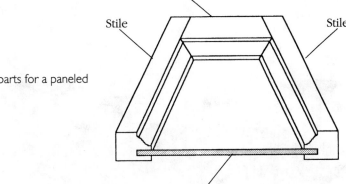

Rail

Stile   Stile

3-28 Frame parts for a paneled door.

Panel

for this work, they can hardly be considered "special," but they might be viewed as special by the dedicated woodworker who is confronted with a one-time counter-covering job.

The same thought is not so extreme when slotting cutters are involved. For example, the countertop fabricator no doubt feels they are necessary to easily form grooves in countertop edges that will be covered with a press-in metal molding and a woodworker can make use of them in off-beat fashion in some joint designs. Workers making many projects like picture frames and

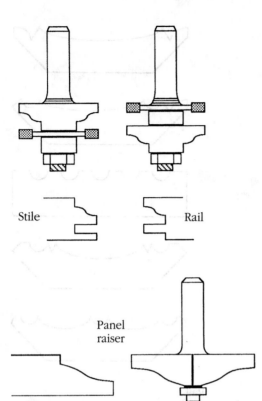

Stile

Rail

Panel
raiser

3-29 Reversible stile and rail assembly from Byrom is usable on ¾-inch to 1-inch stock. The cutters, which should be used in a router/shaper table, are available in various profiles.

3-30 The cutters in The Door Shop set offered by Zac Products, Inc., include ogee, slot-cutting, and panel-raiser bits. All have tungsten carbide blades and must be used in a heavy-duty, ½-inch-capacity router that is mounted in a router/shaper.

plaques consider the key slot cutter essential, while others not so involved might make do with a different hanging method for their pictures or plaques.

The router bits shown in FIGS. 3-32 through 3-47 can be considered special or routine. All are interesting, and anyone owning a portable router should be aware of them.

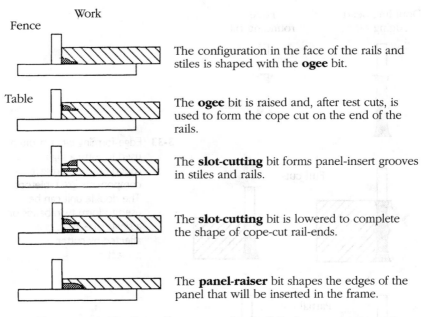

Fence Work

The configuration in the face of the rails and stiles is shaped with the **ogee** bit.

Table

The **ogee** bit is raised and, after test cuts, is used to form the cope cut on the end of the rails.

The **slot-cutting** bit forms panel-insert grooves in stiles and rails.

The **slot-cutting** bit is lowered to complete the shape of cope-cut rail-ends.

The **panel-raiser** bit shapes the edges of the panel that will be inserted in the frame.

3-31 The cutters in The Door Shop set provide flexibility because cutting setups can be organized in relation to the thickness of the stock.

3-32 Half-round bits are often called bull nose bits, probably because large ones can be used to form the shape (bull nose) that is often found on the front edge of stair treads. The radius on bits of this type can range from $\frac{3}{32}$ to $\frac{5}{8}$ inch.

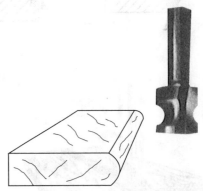

GETTING MORE FROM A BIT

In terms of shapes that can be formed with a single cutter, many bits are multipurpose. When making a full cut, the shape produced is the reverse of the cutter's profile. But with adjustable actions, such as cutter projection (depth of cut) and the width of the cut that can be regulated with guides, you can achieve partial cuts that, in effect, are variations of the full design (FIG. 3-48).

Other examples of bit technique are shown in FIGS. 3-49 and 3-50. In FIG. 3-49, the cut made on one edge is repeated on the opposite edge. Figure 3-50 shows a cross section of the result when one cut is repeated on all edges of the material. These few ideas are offered merely to spur you into

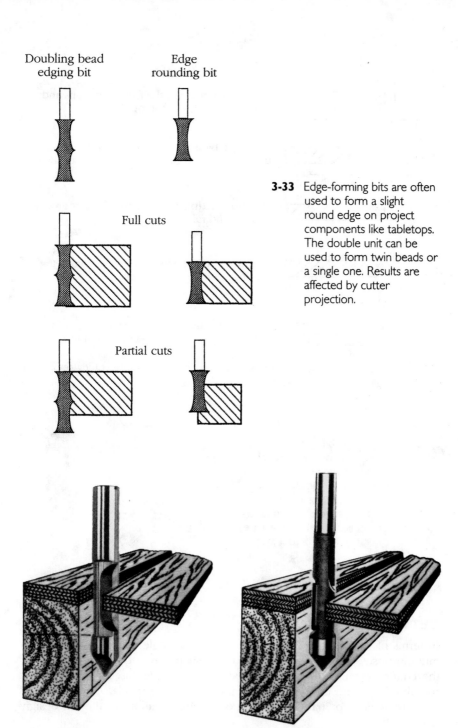

Doubling bead
edging bit

Edge
rounding bit

Full cuts

Partial cuts

3-33 Edge-forming bits are often used to form a slight round edge on project components like tabletops. The double unit can be used to form twin beads or a single one. Results are affected by cutter projection.

3-34 With pilot panel bits you can plunge through a workpiece at any point. A common application is making a cutout for a sink. Other uses include trimming plastic laminates and wood veneers and making cutouts in wall paneling for switch boxes and the like. The one on the right is a "stagger tooth" design.

3-35 Examples of bits that are specially designed for trimming plastic laminates. The one on the left is a pilot panel bit. The short one is solid carbide. The remaining ball-bearing piloted ones have carbide cutting blades. The bit on the right forms a 22-degree beveled edge, which is often the way the edge of a laminate-covered countertop is finished.

3-36 Flush trimmers work precisely because the diameter of the pilot and the cutter are the same. The example on the right has a pilot between two cutters, thus the bit trims top and bottom cover material simultaneously.

thinking imaginatively. Using every bit exactly as it was designed places limits on router versatility. In my shop I keep full-size profiles of all cutters on tracing paper. Then, over an actual-size drawing of the edge I will work on, I can preview what can be done or test the practicality of what I envision.

CARE OF BITS

An essential part of portable router craftsmanship is maintaining bits in pristine condition. Part of this is establishing a safe storage environment. Cutting edges on bits, especially carbide-tipped units, are easily nicked, and

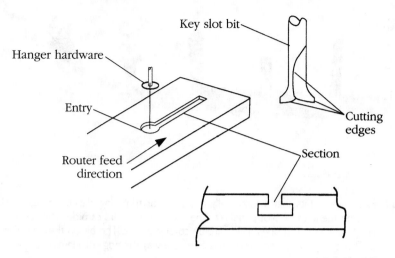

3-37 The key slot cutter makes a limited-depth plunge cut that is then extended to form a T-slot. It's a very practical bit for forming the hanger shape that is commonly used on the back of plaques, frames, and other wall-hung projects.

3-38 Slotting cutters are usually assemblies that include an arbor, cutter, and pilot. Arbors are available in ¼- and ½-inch shank diameters; cutters for various slot widths are interchangeable. The bit is popular for cutting slots in countertop edges that will be covered with a press-in, metal molding. Cutting grooves for splines and feathers in joinery work is another application.

result in an unwelcome volunteer detail on the cuffer's profile. Storing bits in a drawer is not good practice.

An easy and practical storage system, however, consists of a 1½-inch-thick block of wood that has been drilled to receive the shanks of the bits. Don't just drill equally spaced holes, plan the hole spacing to accommodate the various cutting diameters of the bits. Also, allow sufficient finger room between the bits for easy and safe removal of a bit.

3-39 Bowl Bits, from Nordic, are available with cutting diameters ranging from ⁷⁄₁₆ inch to 1¼ inch. They maintain the corner radius while producing a flat bottom.

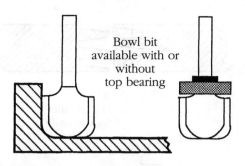

Bowl bit available with or without top bearing

3-40 The Classic Multi-Form bit (Byrom), which should be used in a router/shaper table, can be used for more than 40 different profiles and molding shapes. Unlimited possibilities are available with different combinations of cutter height and fence position.

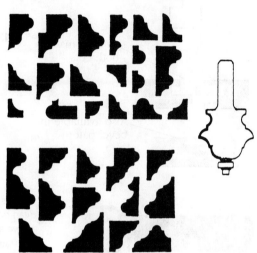

3-41 Finger Joint bits are available with a set number of integral blades or with removable blades. It's easier to set the bit for different stock thicknesses when the number of blades can be arbitrary.

Original containers can be used to store bits, especially those supplied in sets and which are packaged in neat plastic or wood chests. Even so, to avoid the possibility that the cutters might move about and bang against each other, stuff the open areas in the containers with lintfree cloth before putting the cutters away.

Clean the bits, as well as the pilots and ball-bearing guides, after each use. Adhesive materials, such as resin tars and pitch, can accumulate on a bit to the point where chip clearance is reduced and friction is increased. Both factors contribute to poor quality cuts and extra strain on the router and you. Special wood-pitch removers and other solvents are available for

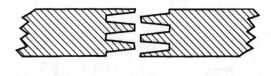

3-42 Finger Joint bits make super strong end-to-end connections, and can also be used in edge-to-edge joints and for miters.

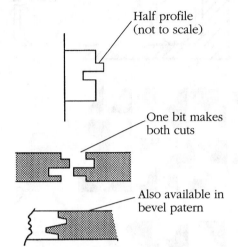

Half profile (not to scale)

One bit makes both cuts

3-43 The single glue joint bit is used to form both parts of the connection.

Also available in bevel patern

3-44 Sometimes called a "dishing bit," this unit can be used for straight grooves that will have a flat bottom and rounded inside corners. Cutting diameters can range from $\frac{7}{16}$ inch to $1\frac{1}{4}$ inch. The bit can also be used for bowls and trays.

3-45 With this dado clean-out bit, the edge of the ball-bearing rides on the shoulder of the dado (or rabbet) so the cutter cleans out any irregularities in the bottom of the cut. It is available in various diameters.

3-46 Finger pull bits are available in various shapes and sizes.

3-47 Two examples of the finger pull forms that can be produced with finger pull bits.

Bit

Bit

3-48 Variations in cutter projection and the width of cut determine the shape of your end product.

Bit

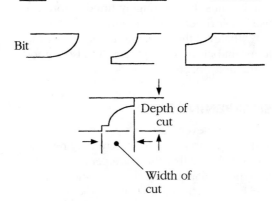

Depth of cut

Width of cut

3-49 Similar cuts made on each edge of workpieces result in batten-type strips.

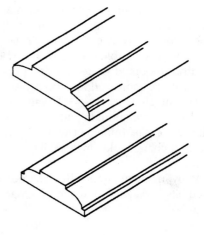

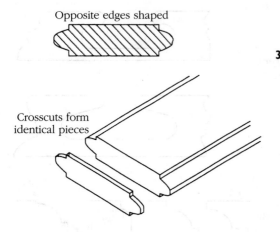

Opposite edges shaped

Crosscuts form
identical pieces

3-50 This is a cross section of stock with all edges shaped. When a long workpiece is edge-shaped this way, it can be cross-cut into narrow strips to produce many similar pieces.

cleaning router bits and other woodworking tools. Be sure to read the safety instructions on the containers of such materials before using them. Often I find that just a gentle scrubbing with an old toothbrush, warm water, and detergent is enough to do the job and is better than working with a solvent. The bit must be thoroughly dried before it is put away. When necessary, I use a hair dryer or a heat gun.

Whatever the cleaning method, coat the bits with a light oil or a spray lubricant before storing them. It's also a good idea to use the same lubricant in the holes of storage blocks.

SHARPENING BITS

Router bits, even carbide-tipped ones, do get dull and require sharpening and will get to the point where they need to be renewed or discarded. I'm a firm believer that there are certain functions in tool use that are best left to experts. Sharpening woodcutting tools, especially router bits with their special characteristics, is one of them.

Particular HSS bits, like those with straight flutes, V-cutters, and the like, can be touched up with India or aluminum oxide slip stones. Work on the inside of the cutting faces while maintaining correct rake angles. The same kind of work can be done on carbide-tipped bits with a 400- or 600-grit diamond hone. Advance the thought to grinding instead of light touch-up work, and it is feasible for the work to be done with power, say with a suitable arbor-mounted stone in a drill press.

For those more serious about bit sharpening, there is a router cutter-grinding attachment (FIG. 3-51) available.

The amateur shouldn't attempt grinding or even touch-up work on the perimeter of bits. Imperfect work can reduce cutting diameter, affect clearance, and change rake angles, all of which contribute to inferior bit performance.

Many manufacturers offer sharpening services for HSS and carbide-tipped bits and can also, when necessary, replace carbide blades. Some will even shape a bit to your specifications. Chances are that there are local experts that can help with your sharpening needs so you don't have to work by mail. Overall, in the interest of quality work with minimum hassle, accept the contribution that an expert sharpener can make.

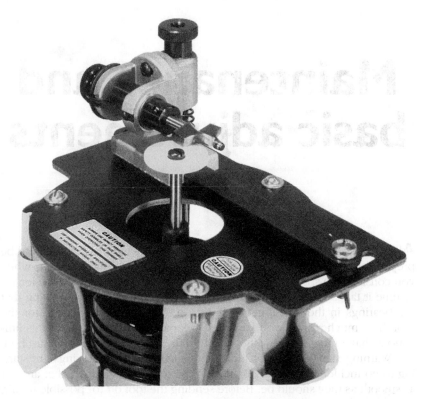

3-51 If you want to attempt to grind your own router bits, it's probably wise to acquire an accessory like this Sears Cutter Grinding Attachment. Special grinding wheels are available for both steel bits and for carbide-tipped bits.

4

Maintenance and basic adjustments

All power tools eventually manifest symptoms of wear that users should be prepared to diagnose—the portable router is not exempted. In fact, when you consider the speed at which the average router operates and the length of time it takes to get through many routine operations, it's surprising that the bearings in the tool don't malfunction sooner. How you care for the tool, how much respect you have for its capabilities, and especially its limitations, have much to do with its useful life before an overhaul is justified.

Warning signs that might indicate that your router's bearings are starting to go include excessive vibration in the tool itself and cuts that are not as smooth as they should be. Before sending the tool off for possible repair, examine the router bits you have been using to be sure that they are in prime condition. A bit with chipped cutting edges, one that is not running concentrically, or one that has been incorrectly ground can cause abnormal vibration and rough cuts. Check the cutter for straightness. It's not likely that a bit will bend, but it is a possibility.

If there is nothing wrong with the cutter, then it's time to suspect the bearings. Unplug the tool and remove any mounted bit, the collet locknut, and the collet. Hand-turn the spindle (motor shaft) and attempt to sense any roughness or irregularity. It's also a good idea to try to move the spindle in various radial directions. If the spindle moves in any direction, the tool should be sent to a factory repair shop for inspection and any necessary repair to its bearings.

Collets can wear, and when they do, a bit will *run-out,* which simply means that the bit will rotate erratically. You can check a collet (with tool unplugged) by inserting a length of drill rod or the longest bit you have and finger-tightening the locknut. If you feel movement when you apply some left and right pressure against the test item, replace the collet.

Remember that incorrect operational procedures can cause excessive vibration that might lead you into thinking the router is failing. Typical operator faults include trying to make a cut that is either too deep or too wide for the router in a single pass and trying to speed up an operation by force-feeding the tool.

Keep the router clean. Use a brush or a blower to remove waste from around the collet area, motor, and so on, where it's likely to collect. Check the owner's manual for information about replacing motor brushes. Examine electric cords and plugs occasionally to be sure they are in good condition.

DEPTH AND WIDTH OF CUT

Depth of cut has to do with the bit's penetration; *width of cut* has to do with the distance across a cut (FIG. 4-1). Depth of cut is controlled by how far the bit projects below the subbase of the tool. Width of cut is controlled by pilots or with guides. How deep and how wide you can cut in a single pass depends partially on the cutter, but mostly on the horsepower of the tool. Heavy-duty routers (2- to 3-horsepower) are considerably more tolerant under conditions imposed by deep/wide cuts than light-duty tools, but all routers have limitations. You'll know when you are abusing the tool and asking for trouble in the cut because the tool will slow up excessively, overheat, and make an abnormal noise. It's always best to reassess the pass any time you must apply excessive feed pressure to keep the router moving.

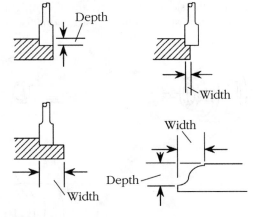

4-1 The depth of cuts and the width of cuts are two basic, operational router adjustments.

Efficient router use calls for the right combination of feed pressure and depth and width of cut in relation to the horsepower of the tool and the material on which you are working. Router expertise naturally comes from obeying the rules and allowing the tool to work *with* you.

COLLETS

Collets come in different sizes to accommodate bits of various shank diameters (FIG. 4-2). Light-duty routers are usually limited to a ¼-inch collet, while

4-2 Collets are available in different sizes. The collet on the left is designed to accept adapter sleeves. This type of accessory might not be suitable for use in all collets.

larger routers that typically use a ½-inch collet can operate with smaller diameter ones. Typical collet sizes include ¼, ⅜, and ½ inch. Being able to handle bit shank diameters of various sizes increases the versatility of any router.

Some manufacturers' router designs accommodate different shank diameters by use of adapter sleeves that slip into the standard collet (FIG. 4-3). Sleeves are provided with the tool or are available as accessories. The design of the sleeves can vary, and they might not be interchangeable among tools, so be sure to check instructions for correct installation.

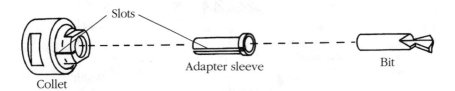

Slots

Collet   Adapter sleeve   Bit

4-3 Adapter sleeves make it possible to use bits of various shank diameters in a single collet. Be sure the slot in the sleeve lines up with the slot in the collet.

Treat collets and locknuts as carefully as you would any precision instrument. Check them occasionally for wear and keep them clean and polished. If you have several, coat the idle ones with a spray lubricant and store them safely.

SECURING BITS

Many routers require two wrenches to secure their bit in place (FIG. 4-4). One wrench is used to keep the motor spindle from turning, the other tightens the locknut. How to go about mounting a bit or changing from one to another has much to do with personal preference. Wrench-turning room, when the router base is in place, is pretty tight, so some operators remove it until the bit is secure. The router can be placed on its side or, if designed to permit it, stood upside down.

4-4 Router bits can be secured with one or two wrenches, depending upon whether the particular router has a built-in spindle lock or whether it must be hand-locked.

A spindle lock is available on some units to assist with bit-changing procedures. The spindle lock keeps the spindle from turning so that only one wrench, used on the collet locknut, is required.

Insert bits fully into the collet and then retract them not more than ⅛ inch before tightening the locknut. Don't attempt to increase depth of cut by minimizing the length of shank the collet will grip. Also, don't test your strength when tightening the nut. Overtightening can cause problems when you wish to remove the bit and can damage the collet. Tighten only as much as is necessary to keep the bit from slipping in the collet when you are cutting.

DEPTH OF CUT ADJUSTMENTS

The base of the router is a sleeve that incorporates a system for vertical adjustment of the motor. Being able to raise or lower the motor in the base is what accounts for the distance the bit projects beneath the base (depth of cut).

A common system has mating spiral grooves on the motor and on either the inside surface of the base or on a pin in the base. Either way, the motor is raised or lowered by turning it like a screw. In addition, there is a depth adjusting ring, usually marked in graduations, on the top perimeter of the base. Typically, to use the ring you need to rest the router on a piece of wood and adjust the motor until the installed bit just touches the wood. With the base secured, just turn the adjusting ring until its zero mark is opposite the index mark that will be on the motor housing to loosen it. Then, tilt the router and adjust the vertical position of the motor until the index mark on the motor housing reaches the desired setting on the ring—your desired cut depth. The procedure might sound complicated, but it's pretty straightforward when you are actually doing it. The cutter's projection can be adjusted by simply measuring with a flex tape, ruler, or a more accurate depth gauge.

Leichtung Workshops offers a handy router depth gauge that's clearly marked in 16ths of an inch from ⅛ inch through 1 inch (FIG. 4-5). The gauge, made of aluminum, is usable for 15 different depth settings.

4-5 This router depth gauge has ¹⁄₁₆-inch graduations and can be used for fifteen exact settings.

Another way to go is to make your own gauge, like the one in use in FIG. 4-6. Since the tool sits flat on the router's base, it's easy to adjust projection while you read the scale. Or, you can preset the gauge for the projection you need and then adjust the bit accordingly. The gauge can be used on other shop tools, such as for setting the height of a table saw blade. Figure 4-7 provides construction details for this depth gauge.

Some routers employ a rack-and-pinion system to adjust the motor's height in the base (FIG. 4-8). It's a neat design that might even include a rotating scale so you can read settings. In the interest of accuracy, however, it always pays to actually measure the cutter's projection regardless of how the system works.

Depth-of-cut adjustments are made with the motor loose in the base. Always remember to use the base locking mechanism before you start working (FIG. 4-9).

Depth-of-cut systems on plunge routers are a little different. Arrangements can consist of a stop rod, a scale, and possibly, a rotatable stop block that

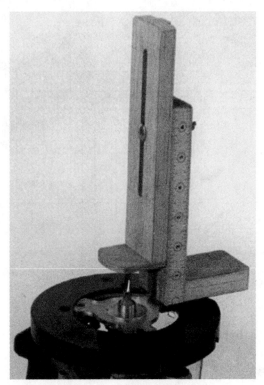

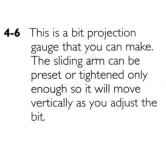

4-6 This is a bit projection gauge that you can make. The sliding arm can be preset or tightened only enough so it will move vertically as you adjust the bit.

makes it possible to preset for three cut depths (FIG. 4-10). With this system, the first step is to release the stop rod so its bottom end contacts the stop block. The motor, which is spring-loaded so it is normally in an "up" position, is pushed down and locked by whatever means the router employs so that the installed bit touches the surface of the workpiece. This establishes the zero setting. The next step is to adjust the stop rod and lock it when its pointer is on the scale graduation that indicates the depth of cut. Thereafter, the bit projection will be to the predetermined depth each time you press down on the motor. By adjusting the three screws that are in the turret-type stop block to different heights, you can preset three different depths of cut.

The option of presetting depths of cut is a great feature of plunge routers, but not mandatory for their use. All you have to do anytime you choose to work without the feature is to lock the stop rod in an elevated position.

PILOT BEARINGS LUBRICANT

Ball-bearing pilots on router bits are durable, but they last even longer and function better when they are treated occasionally with a specially formulated, nonclogging lubricant that helps the bearings stand up to the friction and temperatures that are generated by the high-speed routing. The synthetic lubricant is available in applicators with a needle tip that lets you apply a drop exactly where it's needed (FIG. 4-11).

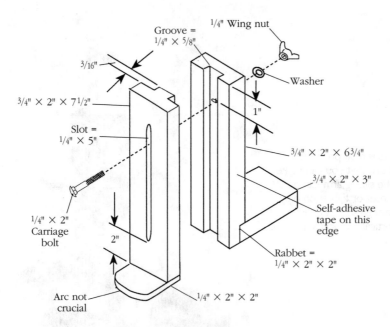

4-7 Construction details for the depth gauge.

4-8 The motor in this Black & Decker router is raised or lowered in the base by means of a rack-and-pinion mechanism. The wing nut at the left is used to secure the motor at the height needed to achieve the depth of cut.

4-9 All depth-of-cut adjustments on conventional routers are made with the motor free to move vertically in the base. Be sure to use the locking mechanism after you are satisfied that the cutter projection is correct.

4-10 Typical depth-of-cut arrangement on a plunge router.

4-11 Special pilot bearing lubricant comes in tubes with a needle point that makes it easy to be precise with the application.

5

Basic tool handling

The portable router is a tool that will do, or attempt to do, whatever you ask of it. Recognizing that the tool and the operator must work together as a team, is the first and major step toward working efficiently with a router or any tool. Like any team assembled for the first time, coach and players should get to know each other and learn to cooperate by going through some practice sessions.

If having a portable router is a novel experience, the first step is getting a "feel" for the tool. It's assumed that you have studied the owner's manual and know the particular characteristics of the machine and have followed instructions for adjusting the motor in the base, securing a bit, and so on. Now, while holding the tool firmly, turn it on for a second or two and then turn it off. Do this several times so you can experience the characteristic starting torque and be prepared for it. Be aware that when you shut down the tool, the bit continues to turn for a while. Whether you continue to hand-hold the tool or set it on its side while holding it firmly until the bit stops is unimportant. Chances are that on particular occasions, you will opt for one of these methods. Just remember, in the interest of safety, avoid having the bit touch you or any other object while it is still spinning.

For cutting practice, have a few reasonably sized pieces of soft wood, like pine, on hand. Be sure the edges you will work on are smooth and straight. For starters, plan to make a shallow cut with a decorative, piloted bit.

BIT ROTATION AND FEED DIRECTION

If you examine the router head on, you will note that the motor, and thus the bit, rotate in a clockwise direction (FIG. 5-1). The ideal course (*feed di-*

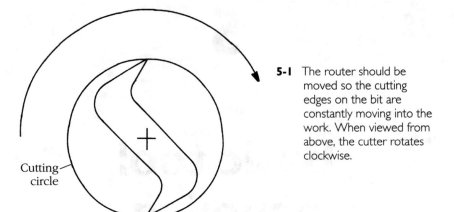

5-1 The router should be moved so the cutting edges on the bit are constantly moving into the work. When viewed from above, the cutter rotates clockwise.

Cutting circle

rection) for the tool when making a cut opposes the motor's torque and keeps the cutting edges of the bit moving *into* the work. On a straight cut, the feed direction is from left to right. Moving the router in the opposite direction gives the cutter a chance to move like a wheel along the work edge, and the operation requires considerably more control. It's also likely that the cut will be rougher than it should be.

Follow the sequence of passes shown in FIG. 5-2 when it is necessary to shape all edges of the work. Cuts that are made across end grain almost inevitably result in some splintering or feathering where the cutter leaves the work. The passes that are made parallel to the grain of the wood remove these imperfections. Also notice in FIG. 5-2 that while feed direction is a normal left to right on the outside edges, the edges of an internal cutout are shaped by moving the router in a clockwise direction.

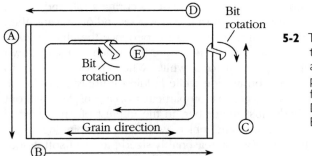

Bit rotation

Bit rotation

Grain direction

5-2 The correct procedure to follow when routing all edges on a workpiece. Do A and C first, or follow A, B, C, D in sequence. Follow E for inside cuts.

The same techniques apply when shaping curved or circular edges (FIG. 5-3). Move the tool counterclockwise (left to right) on outside edges and clockwise when shaping inside curves or circular cutouts.

There are times when the feed direction rule might be broken. This will happen, for example, when doing freehand routing to recess a background. There will also be situations where it is practical and expedient to use several feed directions. Sometimes the router must be moved obliquely across

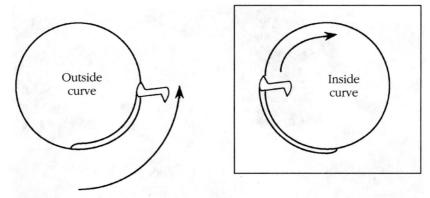

5-3 Ideal feed directions when shaping edges of circular work and curved pieces. For outside curves (A), move counterclockwise. For inside curves (B), move clockwise.

wood grain. The latter situation and some others still permit a reasonable feed direction, but the cutter will often encounter strange grain patterns. Just remember, always control the tool so that constant contact is maintained between cutter and work, and keep an appropriate feed speed.

Feed speed is variable. In a factory where quantities of similar parts are routinely spewed out, setups are established so the rate at which the work moves past a cutter is mechanically controlled for compatibility with the rpm of the cutter. Factors considered by tool setup engineers are denseness of the material, design and size of the cutter, depth and width of the cut, and so on. Regardless of paper work technology, it's results that count. The ideal solution might come to light by means of trial and error.

Somewhere between forcing the router to do more than it can and being too cautious is the ideal feed speed for the work at hand. It's a judgment that becomes easier to wisely make as you progress with router work. If you feed too fast, the motor will complain. If you feed too slowly, you won't accomplish as much as you can and might hold contact between bit and wood in a given area long enough to generate excessive heat, which can burn the wood and even draw temper from the bit. Ideally, you will move the router calmly without excessive pressure while keeping the cutter working.

Important factors include the horsepower of the tool, the condition of the bit, the size of the bit, the depth and width of the cut, the density of the material, and grain direction. As you get deeper into portable routing, these considerations become part of a sixth sense that helps you get through all situations with minimum fuss and good results.

Start all cuts with the router firmly in position but with the bit away from the work edge. Make initial contact slowly, *after* the tool has been switched on and has attained full speed. The direction of feed depends to a good degree on the job being done and the size of the work. Remember that, whenever possible, the cutter should be moving into the work; you can push the router, pull it, or take a frontal position and walk with it as it moves through its cut (FIG. 5-4). Turn the tool off after the cut is complete.

5-4 The operator's convenience is a prime factor when determining whether to push the router, to pull it, or to stand in front of it and walk along with it. The latter method works best when working on long boards.

Always be aware of that spinning bit, especially when you are moving the router toward your body.

Because only part of the base is on the work when you are making edge cuts, pay special attention to keeping the router level throughout the pass. Any tilt will cause the bit to dig in and mar the work; it might even jerk the router so you lose your grip. When I can, I place a support piece under that part of the base that isn't on the work. Any straight piece of wood that matches the work's thickness will do (FIG. 5-5). The idea can also serve when shaping curved edges. Often the waste piece that remains after the project component has been sawed to shape can be used as shown in FIG. 5-6. The support pieces make it a lot easier to keep the router level.

KEEPING WORK SECURE

The material on which you are working must stay put during the router operation. This is not a problem when, for example, you are routing details on a large, assembled project (FIG. 5-7) or when you are shaping edges on a large plywood panel. However, special precautions should be taken when the workpiece is small and light enough to move about as the router is applied.

One system that works for individual pieces and that is especially useful when many similar parts are needed is shown in FIG. 5-8. The clamp strips, which are tack-nailed to a bench top or to something like a sheet of

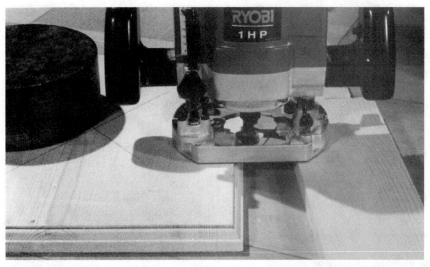

5-5 Since only part of the tool's base is on the work when shaping edges, a support piece under the outboard end of the subbase helps to keep the router level.

5-6 With curved components, the waste that remains when the part is sawed to shape serves nicely as a support piece for the router.

plywood, are situated to accommodate the size and shape of the work. Clamp strips placed against outside edges allow you to shape edges of cutouts. The reverse is also true. When shaping the outside edges on solid material, the clamp strips must be thin enough to allow a cutter's pilot or an edge guide to get by.

5-7 It's often possible to rout decorative details on projects after they have been assembled.

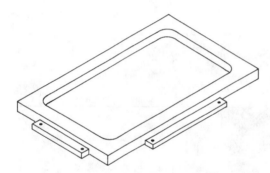

5-8 Clamp strips are tack-nailed on 3 or 4 sides of the work and can be used to keep work from moving as the router is applied.

Tack-nailing, which simply means using the smallest nails or brads that will secure a workpiece to a solid surface, is often used to keep work still. The only objection is the holes that remain when the brads are removed, but they are tiny and can be easily filled. Often tack-nailing can be done in waste areas or in an area that will be concealed by another component, so even that problem is eliminated.

Another idea is to place a heavy steel weight that is part of my workshop equipment on the work to keep it in place while the routing is done (FIG. 5-9).

Pieces too small to tack-nail or secure by other means can be held down with hot-melt glue that is dispensed with a special glue gun like the one in FIG. 5-10. The glue sets quickly and hasn't much penetration so the project will not be difficult to release if the glue is applied in a minimum number of small dots or tiny beads. When ready, lift the project from the surface it's bonded to by carefully using a thin, sharp chisel or knife. Also use the chisel to remove whatever glue that is still adhered to the workpiece. Follow with routine sanding before applying a finish coat.

5-9 A heavy steel weight often serves as a "hold-down" for work being routed.

5-10 Hot-melt glue applied with a special glue gun can temporarily secure small workpieces. The project will not be difficult to release if you sparsely apply the glue.

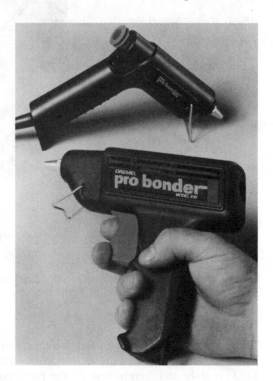

Double-faced tape has many uses in a woodworking shop, among them, temporary work holding chores. There are several types available, but the heavy tape that is used for holding down carpeting should be used for routing jobs. Even small pieces, judiciously placed, will suffice for jobs like keeping workpieces and templates together and securing the work to a temporary base. It's not necessary to overdo with the tape; if you use too much, it may be difficult to separate the pieces.

"MAGNETIC" ROUTER MATS

The new nonslip work mats that are now available from many sources permit freehand routing without the interference of clamps or vises (FIG. 5-11). They are made of various materials but they all keep components in place as you work on them. Some resemble carpet underlayment, while others have an open mesh pattern that helps to minimize dust buildup. Although they are ideal for securing small pieces of work, their size, up to 2 feet × 3 feet, makes them usable for larger projects as well (FIG. 5-12). These mats can also be used to sand small parts.

5-11 Router mats do a good job of securing small workpieces without clamp interference. They are now available from many sources.

SLIM MOLDINGS

The portable router can be used to shape slim moldings, but the best and safest method is to form the shape on material that is large enough to be held securely and then to saw off the part you need (FIG. 5-13). Repeat the procedure when you need many similar pieces, but be sure to sand the sawed edge smooth before you apply the router. When you need many short pieces of molding, it's best to shape the edge of a long board, then cut the strip that you've sawed off into specific lengths.

5-12 The mats are large enough, as much as 2 feet × 3 feet, to secure large parts as well as small ones.

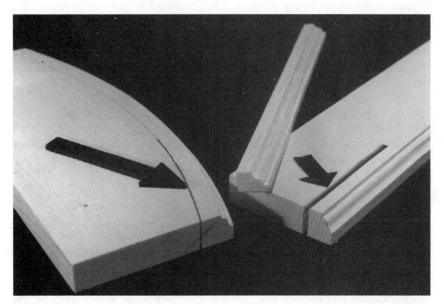

5-13 Pictured is the safe and sane way to produce slim moldings. When all detail is done, simply cut to size. Arrows indicate saw kerfs.

THE FINISH YOU GET

All router cuts should be smooth, but the texture of the cut depends to some degree on the material used (FIG. 5-14). There should be no problem with solid-wood or cabinet-grade plywoods, especially if they have solid

5-14 The quality of cut achieved varies with different material being used.

lumber cores. However, don't expect to get super results on anything like shop-grade fir plywood that, for one thing, often has voids.

Results on other man-made materials, like various types of particle-board, hardboard, and flakeboard, will vary depending on the density of the product. Sometimes routing must be followed with a filler, then sanding before a satisfactory surface for a finish is achieved.

Dense plastics, like a material called Corian (FIG. 5-15), can be routed in fine style, but a tungsten cutter should be used. In fact, it's recommended that carbide cutters be used with most man-made products that use various adhesives or glue as binders to keep wood chips, flakes, laminated layers of wood material, or whatever, together.

A SAMPLE CUT LIBRARY

In my shop, and I'm sure that other router-users do the same thing, I have a "library" of sample cuts like those shown in FIG. 5-16. After establishing and proving a setup for a particular cutter and stock thickness, cut off a short section of the shaped project and store it for use as a gauge when the same operation must be duplicated. This system considerably reduces setup time on future projects and ensures that the correct cutter/wood relationship be established with minimum fuss. You will find that this idea helps whether you are hand-holding the router or using it in a router/shaper table setup.

INCREASING ROUTER SPAN

You will find, in some situations, that the diameter of the router base is not enough to span across support areas. This occurs often when doing recess-

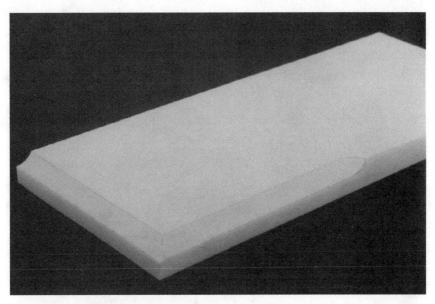

5-15 Plastic materials like Corian can be routed very nicely, but tungsten carbide cutters should be used.

5-16 Sample cuts can be used as gauges to duplicate setups with minimum fuss. It's not a bad idea either to have a set of samples that show the full profile of each bit you own.

ing jobs. The solution is to remove the tool's subbase and to substitute a specially made one (FIG. 5-17). The overall size and shape of the auxiliary subbase is influenced by the job at hand. The location of the attachment holes and the center of the hole through which the bit will pass are determined by using the original subbase as a pattern.

Auxiliary subbases of this type, not necessarily with the shapes that are shown, are often made to accommodate bits with a cutting circle that is larger than the center hole in the original subbase. Two materials that serve well in this capacity are tempered hardboard and a shatter-proof, scratch-resistant plastic, like Lexan.

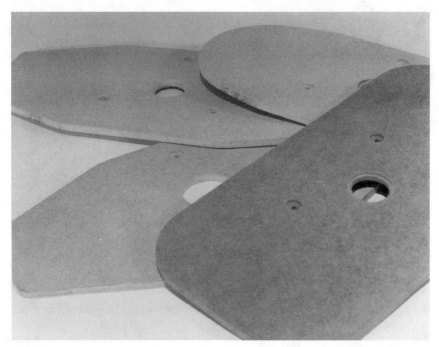

5-17 Special subbases can be made of hardboard or a plastic material like Lexan. With the hardboard subbase, apply several coats of sealer, with a light sanding between them, and then a final coat. Apply paste wax and rub to a polish so the auxiliary base moves easily.

6

Straight cuts

Straight cuts are made along the work's edge or somewhere in the field. They might be parallel to edges, with the grain or against it, across end grain, at right angles to an edge, or directed obliquely across the work. *Through* cuts start and end at edges. A *stopped* cut starts at an edge but is complete before reaching the opposite edge. A *blind* cut starts and stops between edges. Straight cuts can be made freehand by directing the router along a marked line, but this is difficult to accomplish accurately. Whenever possible, guide the router by use of a piloted bit, an edge guide, or by some improvisation that can be as simple as a straight strip of wood clamped in place (FIG. 6-1).

PILOTED BITS

The most straightforward method of guiding a router along an edge is to use a bit with an integral or ball-bearing pilot. You'll get good results if the edge is smooth and free of irregularities and if you keep the router flat on the work and with the pilot bearing against the edge throughout the pass. Some lateral pressure is required for the pilot to do its job, but don't overdo it. If you apply more pressure than is needed, especially with a one-piece bit, the result is either a blemish, like burn marks, or unwanted indents in your project. Such imperfections are less likely to happen with a ball-bearing pilot, but it's still good practice to apply the minimum amount of pressure for the job.

It's often wise to reach the final result by making repeat passes, increasing the depth of cut a little with each pass, with the final pass being just a shaving cut. You'll soon be able to judge when this technique is needed and how to get the job done in a *minimum* number of passes. As always, factors involved are the horsepower of the tool, the size of the cut, and the material.

6-1 A common method of making straight cuts is to guide the router along a strip of wood that is held with clamps.

EDGE GUIDES

Edge guides (FIGS. 6-2 and 6-3) are available for all routers. When one is not supplied with the tool, it should be a first-choice accessory along with a few bits. Buy one that is designed specifically for use with the router you own.

The attachment can be used for cuts on edges that are made with pilotless bits and for cuts like dadoes and grooves that are parallel to an edge. More sophisticated edge guides can be used as shown in FIG. 6-4. After the first cut is made (in this case a dado), the guide is set to establish the distance between cuts.

6-2 The router edge guide is used for cuts on an edge and parallel to the edge, as shown here.

6-3 The guide ensures equal edge distance on all cuts so long as the work edges are smooth and straight. Note that the work is held in a vise, which is a good way to go when work size permits it.

6-4 Some edge guides can be organized to control the distance between cuts.

Straightedge guides, which can be straight strips of wood, tack-nailed or clamped in place, or an on-hand tool, like the clamp shown in FIG. 6-5, are often used in place of commercially made edge guides. A straightedge guide is necessary when the cut is too far from an edge for a ready-made guide to work and when cuts are not parallel to the work edge. When establishing the position of the straightedge, be sure it is parallel to the line of cut and that you have allowed for the distance between the cutting edge of the bit and the outside edge of the subbase.

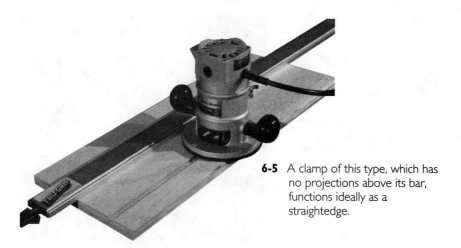

6-5 A clamp of this type, which has no projections above its bar, functions ideally as a straightedge.

It's a good idea to make a note of the bit size and the offset distance you establish for a particular cut so duplicating the operation will be easier. Many professional router users make gauges for future use by noting the bit diameter on strips of wood that have been cut to correct length.

An edge guide idea that is practical for production work is shown in FIG. 6-6. A wood straightedge of suitable length is attached to the subbase with screws that pass through existing holes or even through special holes that you drill. Another way is to make a special subbase with a permanently attached straightedge. Once made, you can use the tool to duplicate cuts exactly at any time. Be sure that the distance between the cutting edge of the bit and the bearing edge of the straightedge is exact (FIG. 6-7).

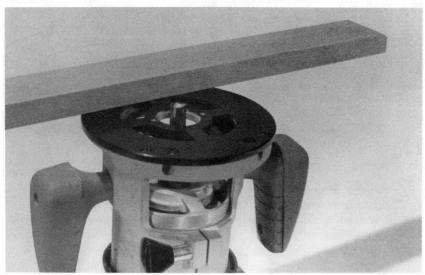

6-6 Production workers often attach a guide of this type permanently to a homemade subbase or to an extra commercial one.

6-7 The distance between the bearing edge of the guide and the perimeter of the bit must equal the edge distance of the cut.

HOMEMADE GUIDES

T-square-type guides, shown in FIG. 6-8 and detailed in FIG. 6-9, are easy to make and will help you achieve accuracy with many router cuts. An advantage is that they are always on hand and often relieve you of the chore of having to improvise with a strip of wood and clamps. The sizes that are suggested work out nicely for average size boards and can usually be secured with a single clamp. If you make larger ones, use wider material for both the head and the blade. Keep the guide secure by using a clamp at its head and a second one at the free end of the blade.

6-8 Examples of homemade guides; one for square cuts, the other for cuts at a 45-degree angle. Assemble the parts with screws or with glue and nails.

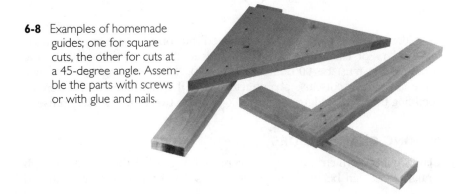

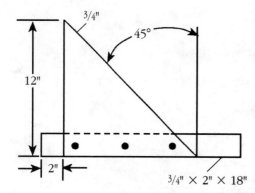

3/4"

45°

12"

2"

3/4" × 2" × 18"

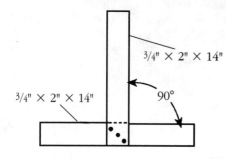

3/4" × 2" × 14"

3/4" × 2" × 14"

90°

6-9 Construction details for homemade guides.

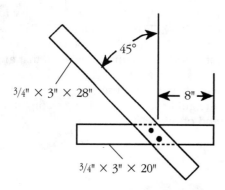

45°

3/4" × 3" × 28"

8"

3/4" × 3" × 20"

Constructing a guide isn't complicated, but obviously it must be made accurately. For example, the angle between the blade and head of the square guide must be 90 degrees (FIG. 6-10). When making one that will serve for angular cuts, use a protractor to set the blade correctly when assembling the pieces. Figures 6-11 and 6-12 show two of the guides in use.

MAKING CLAMP GUIDES

Clamp guides, which are very convenient for guiding the router through straight cuts, can be made in several ways. Two types that are standard

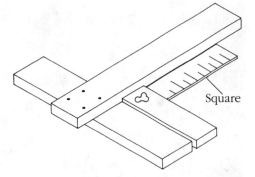

6-10 Guides are simple tools, but don't be casual when making them.

Square

6-11 A clamp should always be used to secure the guide when it is in use.

equipment in my shop, and which you can duplicate, are displayed in FIG. 6-13. Figure 6-14 offers the construction information for one of the clamp guides with a fixed length that is suitable for boards not much more than standard width. An alternate design that suits the clamp for various widths beyond the limits of the fixed version is also shown. What makes the capacity of the extended length design variable is a rear block that can be moved to and fro along a slot that is cut into the fixture's arm (FIG. 6-15). What makes it possible for the clamp to grip securely across the work is the press screw that is mounted in the block at the front end of the clamp (FIG. 6-16).

6-12 Follow the same procedure for cuts made at an angle as for those that are straight cuts.

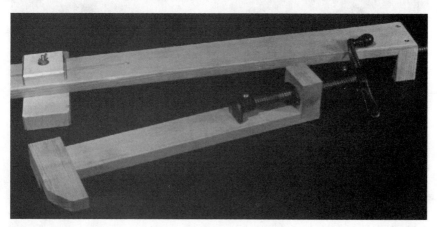

6-13 Homemade clamp guides are fine for straight cuts because they can be locked securely without the fuss of C-clamps.

When organizing the adjustable clamp for use, retract the press screw and lock the rear stop block so the distance between them is an inch or so more than the width of the workpiece. Then, after positioning the guide arm for the cut that is needed, tighten the press screw.

MAKING A FRAME GUIDE

The unit that is displayed in FIG. 6-17, and detailed in FIG. 6-18, is called a frame guide. It has two parallel straightedges that are spaced to accommodate the diameter of the router's subbase. Many times a jig like this is made

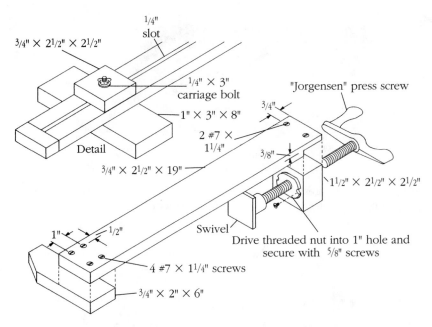

6-14 Construction details for two clamp guides.

6-15 The lock bolt for the adjustable stop slides in a slit that is centered in the clamp guide's bar. Face the underside of the top block with sandpaper so it will grip more securely. You can do this, too, on the area of the stop that contacts the bar.

with fixed blocks at the front and rear. I prefer a fixed front block and an adjustable rear one so the unit can easily conform to the width of the work. It's possible to use the adjustable version without clamp security for some jobs (FIG. 6-19), but not for all work. If only as a safeguard, use at least one clamp at the fixed block end of the jig.

The frame jig can also be used for angular cuts. In such a case, place the adjustable block at the extreme end of the arms and use clamps to secure the jig at the necessary angle. When you make the jig, be sure that the

6-16 The *Jorgensen* press screw is installed this way. Straight grain fir or other wood species like maple or birch are good materials for clamp guides.

6-17 The frame guide has two parallel straightedges that are spaced to accommodate the diameter of the router's subbase.

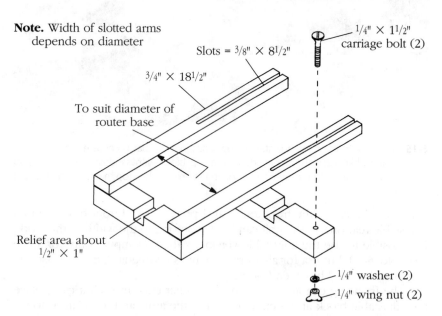

Note. Width of slotted arms depends on diameter

Slots = $3/8" \times 8^{1}/2"$

$3/4" \times 18^{1}/2"$

To suit diameter of router base

$1/4" \times 1^{1}/2"$ carriage bolt (2)

Relief area about $1/2" \times 1"$

$1/4"$ washer (2)

$1/4"$ wing nut (2)

6-18 Construction details for the adjustable frame guide. It's often called a parallel guide.

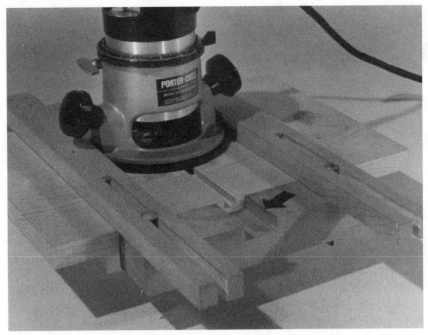

6-19 Relief areas keep the heads on the frame guide from being messed up. Be sure to space the bars so the router can move easily but without side-to-side tolerance.

distance between the arms is just right for the tool. You should be able to move the router without having to force it, but at the same time, there should be no room for the tool to laterally move.

THROUGH CUTS AND BLIND CUTS

A through cut can be one that penetrates the work or, in router applications, one that travels edge-to-edge across the stock as typified by the dado and rabbet cuts shown in FIG. 6-20. The rabbet cut is always on an edge of the stock. The U-shaped cut, called a *dado* when made across the grain and a *groove* when it is in line with the grain, can be made anywhere in the field. These particular shapes will be discussed further in chapter 9, The router as a joinery tool. It's best to guide the router along a straightedge and to work with a straight bit whose diameter matches the width of cut you need. When this isn't possible, the straightedge is repositioned so a second pass opens the cut to the necessary width.

Figure 6-21 shows blind cuts, cuts that start and end in the field with the edges of the work left intact. To make these cuts accurately, stop blocks, which can be secured by some means to the straightedge or to the work itself, are used to position the router at the extremes of the cut. With a conventional router, the tool is tilted and then lowered so the bit penetrates and the router is flat on the work. The tool is then moved along the straightedge until it contacts the second stop block. The plunge router has the distinct

6-20 The dado (U-shaped) and the rabbet (L-shaped) cuts shown here are examples of through cuts. The router is moved completely across the work.

6-21 With blind cuts, stop blocks are used to control the position of the cut and its length.

advantage in this area of portable routing. The plunge router can be set in place, solidly on the work, before the bit is made to penetrate the material by pushing down on the motor.

SURFACE CUTS

Surface cuts are those embellishments that add interest and texture to otherwise flat panels. Just a small example of such cuts are shown in FIGS. 6-22 and 6-23. You do need particular bits to accomplish these cuts, but with a few varieties, your results are limited only by your desire to create.

Except under certain circumstances, decorative surface cuts are accomplished with pilotless bits of the type shown in FIG. 6-24. Many, what you might call *mod* router users, do much work of this type freehand. For starters, though, while still creating impressive geometrical patterns, it's best to make parallel cuts or intersecting cuts by guiding the router along a

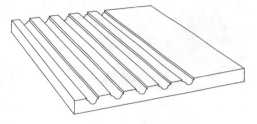

6-22 Surface cuts can be as simple as this: formed in equally spaced, parallel lines. This is also a good way to make molding. The board is cut into individual pieces after the shaping is finished.

6-23 Intersecting cuts add another dimension to surface cutting.

6-24 Surface cutting is usually done with pilotless bits. The depth of the cut, regardless of the cutter you use, has much to do with the effects you create.

straightedge (FIG. 6-25). The results you get will be determined by factors such as the shape of the bit, the depth of the cut, the spacing of the cuts, whether the cuts cross at 90 degrees or some other angle, and so on. It is crucial to work with keen bits because cuts are made across and against the grain. You want the work to be ready for final finishing without tedious touch-up attention.

6-25 Surface cuts turn out best when the router is guided by a straightedge. Mark the workpiece beforehand so you'll know where to situate the guide for each cut. The bit in use here is a V-cutter.

7

Curves and circles

Edges on a workpiece that have uniform or irregular curves can be shaped by moving the router with a piloted bit in routine fashion (FIG. 7-1). As always, a crucial factor for optimum results is that the edge be smooth and free of irregularities.

When shaping edges on circular pieces, work with the same edge guide that is normally used for straightedges (FIG. 7-2). Some of the guides have a V configuration in the bearing edge so the unit can be used without special adjustment. Others are designed for use in a reverse mode that brings into place a bearing edge that conforms more readily to circular edges. Accessories for an edge guide might include arc components that are attached to the straightedge when needed (FIG. 7-3). In any case, while circular edges can be shaped by working only with a piloted bit, you will work more accurately if you utilize any mechanical help that an edge guide might provide.

The bit will encounter various grain patterns when moving along a circular edge and around irregular curves. It will cut with the grain, against the grain, and even quarter the grain, so it's logical to expect that the smoothness of the cut will not be consistent. If you have followed the rules and used a keen bit, however, any extra work should not involve more than a light touch with fine sandpaper on some areas.

PATTERNS

Patterns can be used to guide the router through surface cuts that are not straight. In the examples shown in FIGS. 7-4 and 7-5, the technique is the same as the one used for straight cuts. The guide, in these cases a specially designed pattern, is attached to the work by tack-nailing or with clamps. The location of the guide must allow for the distance between the cutting

7-1 Uniform or irregular curves can be edge shaped by working with a piloted bit and moving the router in a routine counterclockwise direction. Notice how the scrap piece that remains after the part has been sawed to shape is used to help keep the router on a level plane.

edge of the bit and the outside edge of the subbase, which is the same provision required when using a straightedge.

This aspect of portable routing is discussed in greater detail in chapter 8, Working with template guides.

PIVOT GUIDES

Circular surface cuts, which might be simple grooves created with a straight bit or decorative ones with a pilotless bit, can be easily and accurately accomplished by equipping the router so the circular feed path is mechanically controlled. For example, picture a strip of wood with a nail at one end to act as a pivot and a pencil at the other end so you can draw a circle, and then substitute a router bit for the pencil.

The simple arrangement, shown in FIG. 7-6, uses a length of drill rod, drilled at one end for a nail, and secured at the other end to the router base with one of the screws normally used for an edge guide. This is a type of offset arrangement where the radius of the circle is determined by the distance between the pivot nail and the bit, not the working length of the rod.

Some edge guides are designed to accommodate trammel points so that they can be used for pivot cutting, but capacities are within particular limits. It's possible, with some units, to increase capacity by substituting longer rods for the original ones. Because not all edge guides can be used for this

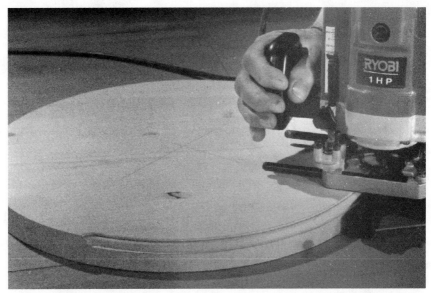

7-2 Some edge guides are designed so they can contribute some extra control when shaping edges of circular pieces. They can also be used on some uniform curves but not on edge shapes, like scallops, that have tight concave and convex lines.

7-3 How an edge guide that is usable on curved or circular edges should be organized depends on its design. This one has extra, arc-shaped components that are screw attached to the basic straight line fence.

particular application, it might be necessary to devise a jig that can guide the router as accurately as any commercial unit.

An example of a homemade pivot guide that I use in my shop is shown in FIG. 7-7. The special ¼-inch plywood base, which is shaped like a paddle,

7-4 It is often necessary to guide the router through a cut by using a pattern. The pattern, like this one, can be used something like a straightedge.

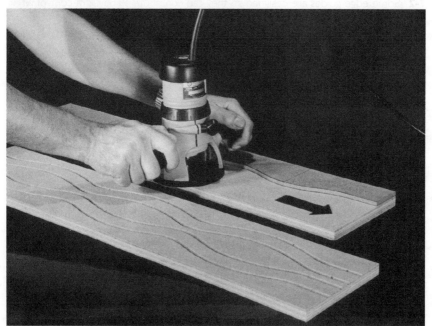

7-5 The pattern can be any shape you wish so long as its contours can be followed by the router's subbase.

substitutes for the regular subbase. Small holes with optional spacing are drilled in the arm of the paddle on a line that passes through the center of the opening for the bit. Make the holes so they will provide a snug fit for a small nail or brad. The length of the jig is optional, but one that will allow

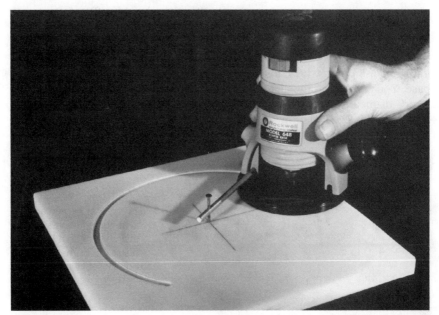

7-6 This is about as basic a pivot guidance system as anyone can devise. The drill rod is the control, but the radius of the circle is the distance from the pivot nail to the cutting edge of the bit.

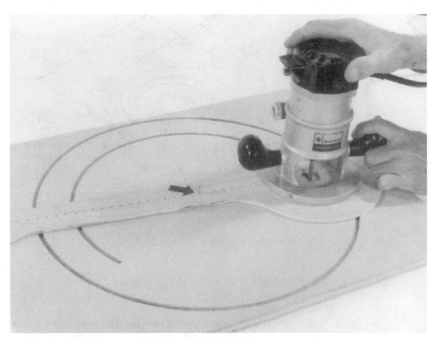

7-7 This homemade pivot guide is actually an auxiliary subbase, but if you choose, it can be attached to the router without removing the regular subbase. The arrow indicates the pivot nail that is at the center of the circular cut.

circular cutting up to about 24 inches in diameter should suffice for average work.

Another type of homemade pivot guide (or circle guide) is shown in FIG. 7-8. This one is more convenient to use because the router merely sits in the retaining ring whose inside diameter matches the diameter of the subbase. You can design the jig for use with several routers that might have different diameter subbases by making several retaining rings that can be attached to the jig's platform with short screws. Construction details for this pivot jig design are shown in FIG. 7-9.

7-8 This type of homemade pivot guide features a retaining ring that is sized to suit the diameter of the router's subbase.

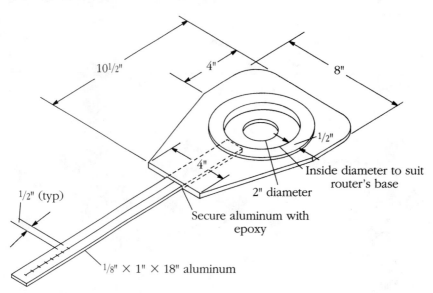

7-9 Construction details for the pivot guide.

There can be times when a commercial jig or one that you have made will not have the capacity for a particular job. In such a case, cut a piece of ¼-inch hardboard or plywood to about 4 inches wide and to your desired length. Draw a centerline on your 4-inch-wide strip, and then, on that line, bore a hole at one end for the bit to pass through. Drive a small nail to serve as a pivot at a point that will provide the radius you need. The temporary jig can be attached to, or used in place of, the subbase.

Pivot guides can also be used for decorative surface cuts (FIG. 7-10). Factors that can contribute to the results include depth of cut, changing circle diameter for repeat cuts, making stopped cuts, and making cuts with different bits.

Pivot guides can be excellent controls when you need perfectly round discs or have to form large holes. For more details on such tasks, see the procedures covered in chapter 11, More router applications.

7-10 Pivot guide jigs can also be used to control the router through circular, decorative cuts.

8

Working with template guides

All router operations are fascinating, but, to me, the most exciting is template routing. This technique can be used for production work or to solve a one-time, difficult woodworking problem. Template routing can be used for incising, for cutting through material and for other tasks that are ordinarily consigned to a jigsaw, saber saw, or band saw.

When the router is equipped with a special guide, it can follow a template to produce projects in quantity, like those shown in FIG. 8-1. Template routing also makes it easy, for example, to form the precise grooves that are needed for sliding and tambour doors (FIGS. 8-2 and 8-3).

Once you get into template routing, you'll discover a host of enjoyable and practical uses for it. If all this makes the technique sound complicated or difficult, don't worry. The most difficult step in template routing is making the template.

Template routing is often called pattern routing too and vice versa. I guess there isn't much wrong with using the words *template* and *pattern* interchangeably, but there is a basic difference between the techniques. When a pattern is used, it is the same size as the part to be duplicated. Cutting is usually done with piloted straight bits or with a panel bit. The pilot rides the edge of the pattern while the bit's blades cut the work in line with the pattern's edge. For example, you might say that the core material for a countertop is the pattern for a plastic laminate cover.

A template is followed by means of a guide that is installed in the base of the router. The guide has a sleeve through which the bit passes. The sleeve, which follows the template, must have wall thickness. This means that the bit is cutting away from the template slightly, so the template is sized a bit larger or smaller than the project, depending on the cut, in order to compensate.

8-1 Special guides that are installed in the router's base, together with homemade templates, allow you to make projects like this in quantity.

8-2 The grooves that are needed for components like tambour doors must be perfectly matched in top and bottom case member. Template guides make it easy to do the work perfectly.

8-3 A template, which is sized and positioned to suit the grooves that are needed, is followed by the sleeve that is part of the template guide.

TEMPLATE GUIDES

The accessory required for your router to do template routing is not an amazing conception but a rather simple, low-cost item. Designs do differ, however, so it is important to buy those that can be mounted in the router you own. Some, like the examples in FIG. 8-4, are attached with screws, while others are assemblies that consist of a threaded body and a knurled, ring-type locknut (FIG. 8-5). All router subbases have a recess so that the only part of the guide to project below the base is the sleeve (FIG. 8-6). This is the part of the guide that bears against the edge of a template. In a sense, it's a pilot; it moves along the template's edge, and so guides the cutter.

8-4 Template guides are not interchangeable among tools made by different manufacturers. Be sure to buy those that are suitable for the router, or routers, that you own. These are attached to the router's base with screws.

8-5 Another type is an assembly that is composed of a threaded base that slips through the hole in the subbase. A ring-type locknut secures it.

8-6 Regardless of its design, the template guide fits into the router's subbase so the only part that protrudes is the sleeve through which the bit passes.

All guides, regardless of how they are secured in the subbase, have common characteristics. One of the most important factors is the inside diameter of the sleeve. If you work with a ¼-inch-diameter bit, then the inside diameter of the sleeve should be compatible and allow the bit to turn without excessive friction. This applies whether you choose to work with a ⅜- or ½-inch bit as well. Another variable factor is the length of the sleeve (FIG. 8-7). In use, there must be clearance between the bottom of the sleeve and the workpiece. If, for example, the sleeve projects ¼ inch below the bottom of the subbase, then the thickness of the template must be greater than ¼ inch. Guides are made with sleeves of different length, but the assortment that is available is not likely to cover all situations. One solution is to use a material for the template that is thick enough to go along with the sleeve length of the guide you wish to use. Another is to shorten, by grinding, the sleeve of a guide so it will work with the thickness of the stock you decide to use for the template.

One way to solve the problem of incompatibility between router bases and template guides is to check out a Universal Base Plate Kit (FIG. 8-8). This kit, available from Leichtung, consists of a special plate that is slotted and drilled so it will fit most any router with up to a 6-inch-diameter round base, four different size template guides, and two centering pins. The centering pins are a wise addition since, by mounting them in the router's collet, you will be certain that the plate and template guides will be centered precisely.

Figure 8-9 shows the major operational factors to consider when preparing to cut with a template guide. Because the bit rotates in the guide's sleeve, and the sleeve walls have cross-sectional thickness, the template is

8-7 The size differences among template guides have to do with the length and the diameter of the sleeve.

8-8 The Universal Base Plate Kit is usable on most routers that have up to a 6-inch-diameter round base. The centering pins assure concentricity of router bits and template guides.

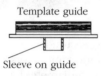

Template guide

Sleeve on guide

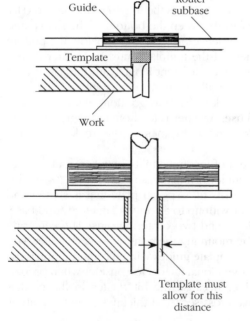

Guide
Bit
Router subbase

Template

Work

8-9 Illustrated here are important factors to remember when making a template.

Template must allow for this distance

sized to compensate for the distance from the cutting edge of the bit to the outside edge of the sleeve. For example, if you are using a guide to form a disc, the diameter of the template will be less than the diameter of the part you need. Likewise, if you are forming a hole, the inside diameter of the template will be greater than the diameter of the hole you need.

Another point to remember is that the intricacy of the template is limited by the diameter of the guide's sleeve. For example, if the sleeve's diameter is ½ inch, it won't be able to get into a corner that has a ⅛-inch radius. Also, because all cutters have a circular cutting path, they can't form a square corner.

TEMPLATES

Some ready-made templates are available. For example, jigs that let you cut dovetail joints are supplied with special finger templates (FIG. 8-10). Others that are available include various sizes and styles of letters and numbers, and those that guide you through forming recesses (mortises) for door hinges. Commercial ones are fine, and it's wise to include some in your workshop equipment, but the real fun of template routing comes with making your own. With your self-created templates, you can design something exclusive for a project, solve a woodworking problem, or conceive a method that makes a routine shop chore easier and more accurate.

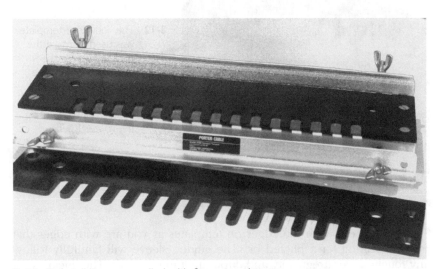

8-10 Dovetail jigs are supplied with finger templates.

Templates can be secured to the workpiece in several ways—by clamping, tack-nailing, or adhering them temporarily with hot-melt glue. Which method you choose depends on factors like the size and shape of the template, whether holes left by tack-nailing are objectionable, and so on. The most convenient method is the best method.

The material you use for templates is optional. Commercial houses that might use a particular template interminably might use aluminum or even steel. Hardboard, preferably tempered, is a good choice for the home shop. Plywood is alright, but edges might become marred after a lot of use. Much depends on frequency of use and anticipated service life. For one-time use, almost any available scrap piece of wood might do—like the example in FIG. 8-11 that served nicely for the decorative cuts on the drawer fronts shown in FIG. 8-12.

8-11 An example of a "quickee" template. Layout and cutting can't be done casually even though you can use almost any piece of scrap wood to make such a one-time-use item. It's probably wise to store any template you make. You never know what the future holds.

8-12 The "quickee" template was used to make the outline cuts for the decorations on these drawer fronts. Backgrounds were recessed a bit by using the router freehand.

Be as careful with the edges on templates as you are with edges that provide bearing for a piloted bit. The guide's sleeve will faithfully follow any roughness or irregularity and duplicate it in the work. The thickness of the template is optional as long as it is compatible with the guide's sleeve length and allows the depth of cut you need. You'll find that ¼-inch thickness works out nicely for most of the work you will do.

DOING THE CUTTING

How to start the cut depends on what you envision as the final result. If the background of the project will be recessed, then the cut can start anywhere with the router moved until it is halted by the contact between guide sleeve and template edge. This also applies if the design will be recessed. In a sense, the template serves as a *stop*.

If the template serves mainly as an outline guide (FIG. 8-13), the sleeve should be in contact with the template and the bit should be turning before it contacts the work. The common solution is to have the tool running while you hold it on the work, tilted just enough to avoid cutter-to-work contact.

8-13 A typical template that is used is an outline guide. The width of the cut depends on the bit you work with. Often cuts like this are made with veining bits, which produce a round bottom groove.

Then, very slowly, bring the router into working position. There are times when you might be able to start with the router level and the sleeve in correct position but without cutter contact. Then it's a matter of lowering the router until it sits solidly on the template. These methods for starting cuts don't apply if you are using a plunge router. With a plunge router, the tool can be firmly in position with the guide sleeve against the template's edge before the bit is lowered into the work.

Once the cut is started, be sure to maintain the necessary contact between sleeve and template throughout the pass. On cuts of this nature, where the bit encounters all sorts of grain patterns and directions, you must be prepared to oppose the router's inclination to take an easier route rather than follow the template. At the end of the cut, lift the router vertically or, as many workers do, turn the tool off while maintaining the end-of-cut position and remove the tool after the bit stops. The whole idea is to avoid marring the work by tilting the router or moving it away from the template.

Outline cuts are often filled with a material that contrasts with the wood, like wood filler or plastic aluminum, as was used on the horse plaque pictured in FIG. 8-14. Designs for the horse template are detailed in FIG. 8-15. It's usually best to apply more filler than is needed. Excess can be sanded flush after the material dries by working with a pad sander or sandpaper wrapped around a block of wood.

8-14 Outline cuts can be filled with a contrasting material. Plastic aluminum was used here, but there are many types of wood fillers that can do very nicely.

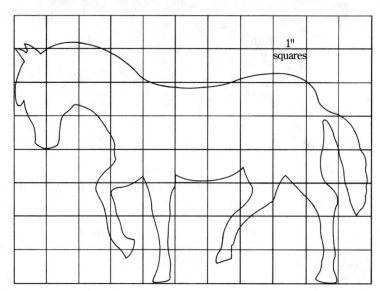

1"
squares

8-15 The horse template can be made as large as you wish by using the traditional squares method depicted here.

TEMPLATES THE EASY WAY

It might not be justifiable to call the setup that is shown in FIG. 8-16 a template; temporary guidance system might be more in order. The point is that elaborate templates that require time and effort to make are not always necessary, especially if what you are doing might never be repeated. While the arrangement might be an improvisation, the results will not lack professionalism. The tack-nailed arrangement of bits and pieces that were shown in FIG. 8-16 proved as efficient as necessary to produce the incised cut shown in FIG. 8-17.

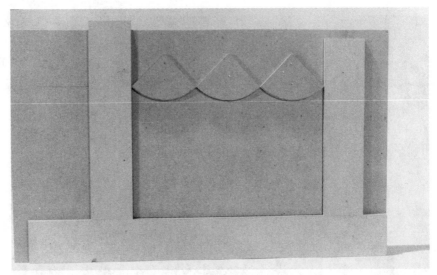

8-16 These are just bits and pieces of hardboard, but when organized and tack-nailed to the work, they serve as well as any one-piece template.

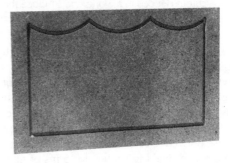

8-17 Improvising templates by using separate pieces of wood doesn't downgrade the results you get. The idea might not be so exciting when you need to repeat the operation many times.

Various cuts can be produced by guiding the router with on-hand straight pieces and discs and squares of wood. For arcs, quarter circles, and scallop designs, a disc can be cut to suit your purposes on a band saw, jigsaw, or with a handsaw. A single template can often be used to create many variations. For example, cuts from the same template can be made in parallel or opposing fashion, or they can cross each other at right angles or obliquely. You do want to give some thought to final results, however. Often the guide can be used to trace lines on paper so you can preview your arrangement.

Figures 8-18 and 8-19 show what might be the extremes of template design. When it was necessary to make cutouts in the top of a workbench, which was a one-time operation, I tack-nailed an arrangement of straight-edges to the bench top to outline the opening I needed. Another shop project involved a set of chairs that were designed with solid wood sides. It was necessary to provide a reuseable template so all the parts (12 were needed) would be exact duplicates.

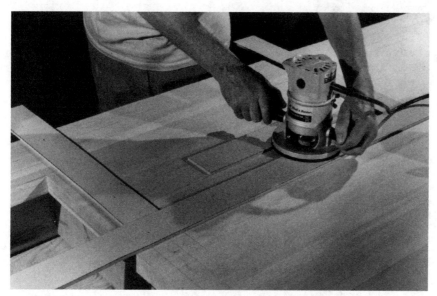

8-18 A series of straightedges were used here as templates to outline the shape of a cutout needed in a workbench. Notice the support piece under the open end of the router base; this helps keep the tool level while cutting.

The procedure required to cut through a workpiece is the same regardless of how the router is being guided. When the material is thin, say ¼-inch plywood or hardboard, it would be a poor router indeed that could not do the job in a single pass (FIG. 8-20). When the material is thick, which is a relative term, the number of passes required, with depth of cut increased for each one, depends primarily on the horsepower of the tool. Judging whether to use one, two, or even three passes relates to how the tool behaves. Smaller routers can rival big routers in terms of getting through particular stock thicknesses. It just takes longer to get there.

A FEW PRODUCTION IDEAS

Most commercial houses design special holding fixtures to make it easier to do repeat design work for components that are required in quantity. There's

8-19 There are many times when improvising a template by using separate pieces just won't do. The part you need might have an intricate profile, or you might need many duplicates. This hardboard template was used to produce 12 side pieces for a set of chairs.

8-20 Templates are also used when it is necessary to cut through a workpiece. Notice how work and template, clamped together, are elevated so the bottom of the bit is in open air.

no reason why, under similar circumstances, you can't do the same. The example shown in FIG. 8-21 is a trough with side pieces separated by the width of the workpiece. The hinged cover, which is actually the template, is swung down after the cut-to-size workpiece is secured in the trough. Repeating the cut along the length of a workpiece is just a matter of repositioning the work for each operation. Marking the work beforehand for the cut spacing you want makes it easy to be accurate each time you move the work.

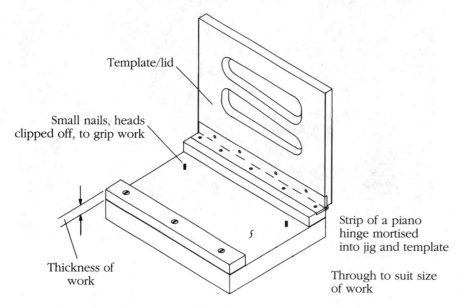

Template/lid

Small nails, heads clipped off, to grip work

Thickness of work

Strip of a piano hinge mortised into jig and template

Through to suit size of work

8-21 Trough-type jigs make it easier to turn out pieces in quantity.

Figure 8-22 shows the details of another version of a trough-type jig. The concept is the same but the lid/template is registered correctly over the work by means of the locating pins. The height of the locating pins should be less than the thickness of the material used for the lid. Jigs of this type can be used for incising surface designs, recessing, and for cutting through materials.

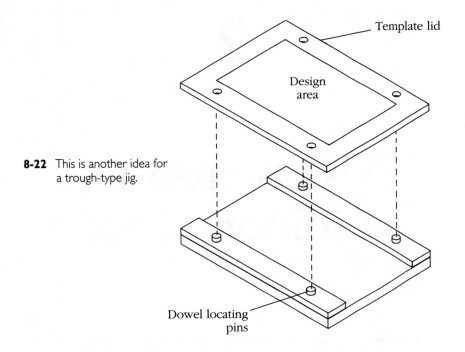

Template lid

Design
area

8-22 This is another idea for
a trough-type jig.

Dowel locating
pins

9

The router as a joinery tool

As you have probably surmised by now, the portable router is usable for more than skin-deep applications like shaping a fancy edge on a tabletop. Fancy edging and the like contribute visual impact and can often be the difference between a so-so project and a professional-looking unit. But it's in the often-concealed areas of furniture where you find the hallmarks of quality construction. For example, the connections between components make the difference between a chair or chest that soon wobbles and one that becomes an heirloom.

Any woodworking joint can be formed with hand tools or with stationary power tools, used individually or in combination. A bit and brace, various chisels, and a few saws are needed to form a mortise and tenon joint by hand. Even with stationary power tools, you would need a drill press for the mortise and a table saw or radial arm saw for the tenon. However, with the unpretentious portable router, the entire job can be done on its own—efficiently, often without having to change the bit, and with reasonable speed. Unless you're a purist with a love of dovetail saws, chisels, and the like, the router is the only way to get precision-fitted, classic dovetail joints in minimum time and with minimum fuss. This observation applies to all the sample joints shown in FIG. 9-1.

It's not all roses, though. The tool does what you ask it to, right or wrong, so careful tool handling is necessary. Most times a guide is required. This might be the pilot on a bit, a simple straightedge, a commercial accessory like a dovetail jig, or jigs that you make. Don't feel that the more advanced joints like the mortise and tenon or dovetail are the only ways to go, even though the portable router makes them more feasible. It's wise to determine the easiest-to-do joint for the project at hand. In other words, you would expect to go all out for long-lived, esthetically pleasing furniture, but

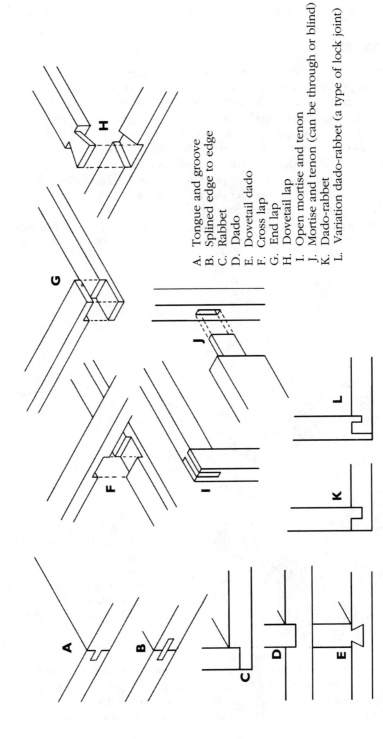

A. Tongue and groove
B. Splined edge to edge
C. Rabbet
D. Dado
E. Dovetail dado
F. Cross lap
G. End lap
H. Dovetail lap
I. Open mortise and tenon
J. Mortise and tenon (can be through or blind)
K. Dado-rabbet
L. Variation dado-rabbet (a type of lock joint)

9-1 Typical joints that can be formed with a portable router. Notice how many of them involve just U-shaped or L-shaped cuts.

the same attention and design values would probably be out of line for shelves in a garage that will only hold garden tools and paint cans.

RABBETS

A rabbet cut is a rabbet cut regardless of whether it is made across the end grain or parallel to an edge (FIG. 9-2). The width of the L-shaped cut usually matches the thickness of the mating piece. The depth of the cut is somewhat arbitrary, ranging from one-half to two-thirds the thickness of the component in which it is made. An extreme case is one involving plywood, where the cut is so deep that only the edge of the surface veneer is visible when the joint is assembled (FIG. 9-3). This is alright as far as appearance is concerned because it conceals unattractive plies in the material, but there will be weakness at the base of the shoulder (where width of cut and depth of cut meet). There is no problem if the project is designed to oppose the lateral forces that might cause a rupture at the weak point. Otherwise, glue blocks or other reinforcements should be introduced to increase the strength of the joint.

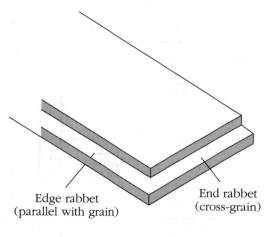

Edge rabbet
(parallel with grain)

End rabbet
(cross-grain)

9-2 The L-shaped cut, called a rabbet, is formed with a straight bit. Use repeat passes when necessary to widen the cut.

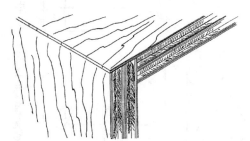

9-3 Rabbet cuts are often deep enough so only the surface veneer of plywood and similar materials is visible. Joints like this should be reinforced with glue blocks when possible.

A common technique that results in smooth, trim corners after the project is assembled is shown in FIG. 9-4. Cut the rabbet a fraction wider than necessary so that after the glue used in the joint is dry, the excess can be sanded off so the edge will be perfectly flush with adjacent surfaces.

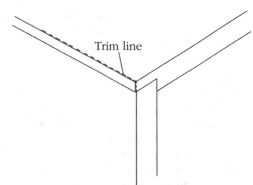

9-4 Making the rabbet a bit wider than the thickness of the mating component is a good idea. The excess is sanded off after the joint is assembled.

Trim line

Rabbet cuts can be formed with a piloted bit, but the maximum width of cut available would be less than one-half the cutting circle because you have to take into account the diameter of the pilot. For example, a rabbeting bit with a 1¼-inch diameter might have a cutting edge (width of cut) of ½ inch and a one-pass depth of cut of ⅜ inch. This is fine for production work when the component that must fit the L-shaped cut is ½-inch thick, but how do you work with the variables that are present in one-time or infrequently confronted situations?

The solution is shown in FIG. 9-5. You work with a straight bit with a cutting circle that comes as close as possible to the rabbet's width of cut, and you guide the router along a straightedge. If the width of the cut you need is greater than the diameter of the bit, then you reposition the straightedge and

9-5 Rabbeting bits are available but are limited in how wide they can cut. Straight pilotless bits, with the router guided by a straightedge, let you cut rabbets of any width.

make another pass. The advantage of a pilotless, straight bit for this kind of work is that you can decide the width of the cut by the location of the straightedge. Any cut width from zero to the real diameter of the bit is available in a single pass.

It's characteristic with the rabbet joint that some of the end grain or the part in which the L-shaped cut is made will be exposed. This can be objectionable. When constructing a case-type project, make the rabbet cut in side members so exposed end grain will be topside and less visible.

Components of case goods and cabinets are often rabbeted along back edges so a panel can be inset as a back seal (FIG. 9-6). When the rabbeting is done before assembly, the panel can have square corners. The corners of the panel must be rounded to mate with the corner radius left by the bit when rabbeting is done after assembly. If the project is something like a kitchen cabinet that will be wall-hung, cut the rabbet anywhere from ¼-inch to ½-inch deeper than necessary so the back edges of the cabinet can be trimmed to conform to any irregularities in the wall.

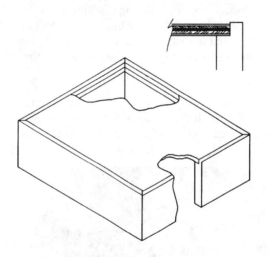

9-6 Back edge of cabinets are often rabbeted so a panel can be inset as a back seal. The rabbeting can be done after assembly or on individual components before parts are put together.

A rabbet is often used in utility drawers as the joint between drawer front and sides (FIG. 9-7). How wide you cut the rabbet determines whether the drawer front will be flush or have a lip. This is not the strongest joint for the purpose, especially if only glue is used, so reinforcements such as nails, screws, or dowel pegs are usually added. These should be driven through the side of the drawer into the shoulder of the rabbet.

DADOES AND GROOVES

Dadoes and grooves (FIG. 9-8) are U-shaped cuts formed in one component to accept the squared end of another part. A typical application is a bookcase where the ends of the shelves are set into dadoes formed in the vertical members of the unit. The width of the cut should match the thickness of the insert piece, but being too precise, to the point where the insert must be forced into place, can cause problems. It's best that the width of the dado

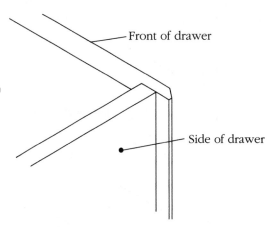

9-7 Rabbets are often used on utility drawers as the joint between front and side pieces. The connection should be reinforced.

Front of drawer

Side of drawer

9-8 The difference between a dado and a groove is the name only. Both are U-shaped cuts; one made with the grain, the other across the grain.

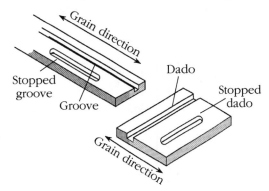

Grain direction

Stopped groove

Groove

Dado

Stopped dado

Grain direction

allow the insert to slip into place, without gaps. The depth of the cut is usually about one-half the stock's thickness. It can be more, but going too deep will create a weakness in the area of the cut.

The advantage of the U-shaped cuts becomes obvious when they are stacked against a butt joint. The shape of the joint forms a ledge for the insert, and there is more surface area for glue. There is a disadvantage though. The U-shape, which is not very attractive, is visible at front edges when the cut runs across the component. When this is bothersome, one of the ideas shown in FIG. 9-9 can be used to conceal the joint or, as they say, "to fool the eye."

Another solution is to make stopped or blind cuts and to shape the insert in one of the ways shown in FIG. 9-10. The only difference between this cutting technique and forming the cut completely across the stock is that blocks are used so the router can be started and stopped at specific points (FIG. 9-11). A plunge router has advantages for this type of operation because the tool can be firmly positioned before the bit is made to penetrate the work. Conducting this procedure with a conventional router calls for tilting the tool to start with and then slanting it down so the bit can get going.

Dadoes and grooves **125**

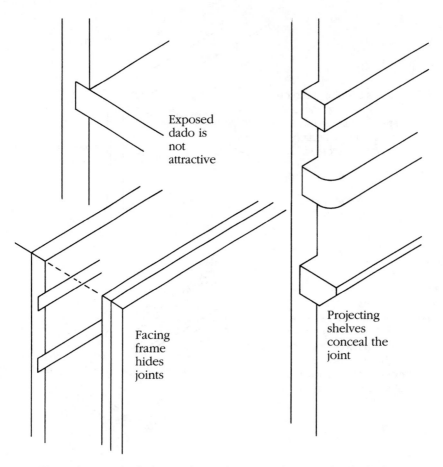

Exposed
dado is
not
attractive

Facing
frame
hides
joints

Projecting
shelves
conceal the
joint

9-9 Pictured are methods that can be used to conceal or to make the U-shaped cut less visible.

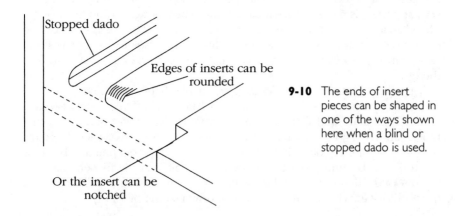

Stopped dado

Edges of inserts can be rounded

Or the insert can be notched

9-10 The ends of insert pieces can be shaped in one of the ways shown here when a blind or stopped dado is used.

9-11 The length of stopped or blind dadoes (or rabbets) is controlled by using stop blocks to determine where the cut starts and ends. It's on jobs like this that the plunge router displays its merits.

Unless the dado or groove you need is reasonably close to an edge so the router can be used with an edge guide, your dadoing and grooving passes will be made using a straightedge. A straightedge can simply be a strip of wood that is clamped or tack-nailed in place. You can also use one of the ideas that were described in chapter 6, Straight cuts, which deals with straightedge use and cutting.

Grooves are often needed in edges of fabricated panels, like plywood or particleboard, so trim can be added to conceal unattractive core material. Often the trim, which is not uncommon on table edges and countertops, is ready-made metal molding with an integral, serrated "tongue" that is forced into the groove (FIG. 9-12). Professionals use slotting cutters for this kind of work, but it's possible, when you lack the tools, to get by with a small, regular straight bit. When using such ready-made moldings, it's crucial that the width of the groove be exactly right for the serrated tongue.

9-12 Grooves for ready-made metal molding are often formed with slotting cutters, but you can get by with a small straight bit when necessary.

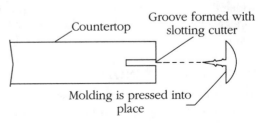

Countertop
Groove formed with slotting cutter
Molding is pressed into place

Figure 9-13 shows other ways to take care of unsightly edges. As you can see, the added strip can be simple or can be designed to contribute to the overall appearance of the project. Strips can be preshaped, but when solid wood edging is used, it's best to do router work after the new pieces are in place. Often a contrasting material is used for the edge strips; for example, strips of walnut or cherry are typically used on the perimeter of a birch or maple table.

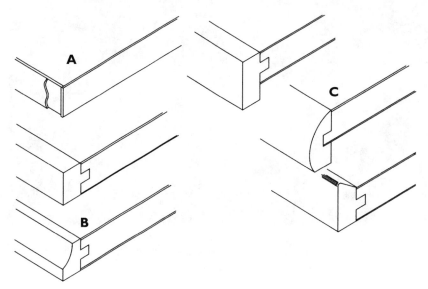

9-13 When concealing edges of panels you can: A. use surface plastic laminate, B. use wood strips (plain or shaped), or C. bulk or lip the edges by using edge strips that are wider than the panel thickness.

HALF-LAP JOINTS

Examples of common half-lap joints are shown in FIGS. 9-14 and 9-15. When the parts to be joined have equal width and thickness, the width of the cut is the same as the stock's width, while the depth of the cut is one-half the material's thickness. This applies whether the joint will be at a midway point or at an end, in which case the joint is technically an *end lap*.

It's always best, whether two pieces or a dozen pieces are involved, to hold the piece together with clamps or some other means, so cutting is done across all of them at the same time. Use a straightedge as a router guide and cut with the largest diameter straight bit you have. Use repeat passes to widen the cut when necessary. End laps should be done the same way. Assuming a four-piece frame, assemble the parts as a unit so you can cut across them as if the assembly were a single board.

MORTISE AND TENON JOINTS

The mortise and tenon joint, whose parts are illustrated in FIG. 9-16, ranks with the dovetail joint in terms of heirloom-quality construction. Often a

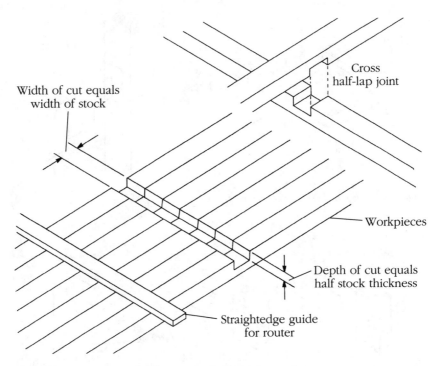

Cross half-lap joint

Width of cut equals width of stock

Workpieces

Depth of cut equals half stock thickness

Straightedge guide for router

9-14 View the cuts needed for crossing half-lap joints as dadoes.

9-15 Cuts required for end laps are just rabbets. The rabbets are equal when the components have the same width.

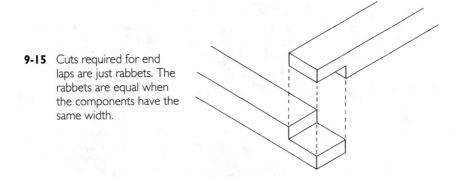

doweled joint is used as a convenient substitute. This is acceptable, but the design isn't more than a butt joint reinforced with pegs. The mating surfaces of the components present end grain to surface grain, which doesn't provide for the strongest glue bond. It's better to arrange for the glue bond to occur long grain to long grain, which is what a tenon in a mortise does. With a mortise and tenon you get maximum glue bond on interior contact surfaces, plus the interlocking feature provided by the joint's design.

There are many variations of the mortise and tenon joint, some of which are illustrated in FIGS. 9-17 and 9-18. The concept is an integral projection on one component that is sized for a cavity, usually rectangular, that is formed in the mating piece. The projection is the tenon, the cavity is the

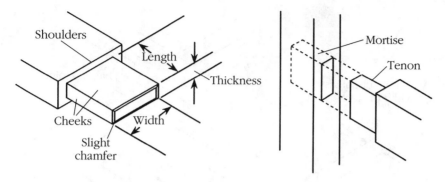

9-16 The typical mortise and tenon joint.

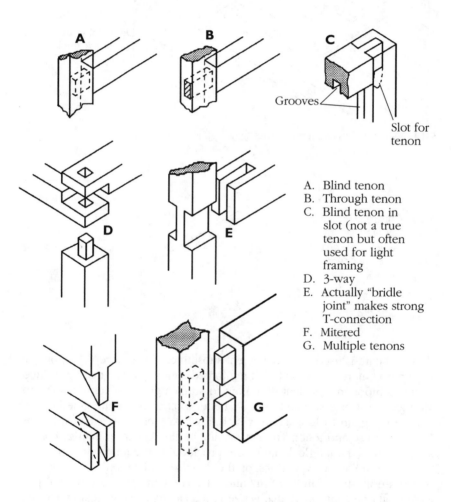

A. Blind tenon
B. Through tenon
C. Blind tenon in slot (not a true tenon but often used for light framing
D. 3-way
E. Actually "bridle joint" makes strong T-connection
F. Mitered
G. Multiple tenons

9-17 Various types of mortise and tenon joints.

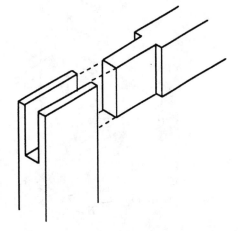

9-18 The open mortise and tenon joint is often used as the corner connection on frames.

mortise. The mortise may be of limited depth (blind) or it can pass through the material. This determines how long the tenon must be. When the tenon is used with a through mortise, cut the tenon a fraction longer than necessary. The amount that projects can then be sanded flush to adjacent surfaces after the glue dries. If the mortise is blind, cut the tenon a fraction shorter than the depth of the mortise; this provides a bit of room for excess glue.

In my shop, the mortise is always cut first. Then the tenons are formed so they slip-fit into the cavity. The joint must not wobble after the tenon is inserted, and conversely, it shouldn't be necessary to use excessive force to seat it either. Tenons that are too tight can cause mating pieces to split, even force excess glue to travel through the pores of the wood. The thickness of the tenon, and so the width of the mortise, should not be greater than one-third to one-half the thickness of the mortised component.

JIGS TO MAKE

Two jigs that I use in my shop that practically eliminate the possibility of human error when forming mortise and tenon joints are a vise clamp (FIG. 9-19) and a special, adjustable router base (FIG. 9-20). One of the problems associated with forming mortises is that they are most often formed in the narrow edge of stock. This poses the problem of keeping the router level as you make the cut. Traditionally, the solution is to clamp an extra piece on each side of the work to provide more support surface for the tool. The vise clamp provides this automatically without the need of additional clamps (FIG. 9-21). The work is gripped by the jaws of the vise clamp that, in turn, is secured in the bench vise. When the same cut is required in several pieces, butt the parts together on a flat surface and hold them with a clamp so you can mark across all of them at the same time.

Figure 9-22 shows the special router subbase ready for mounting, while FIG. 9-23 shows it in action together with the vise clamp. Once the adjustable guides in the subbase are set, you'll know that all the cuts you make will be on the same centerline. Because the guides are individually adjustable, the

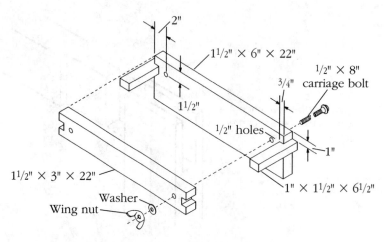

9-19 Construction details of the vise clamp. Drill the holes for the carriage bolts after the parts have been formed and assembled. Be sure the top edges of the jaws are flat and flush with each other.

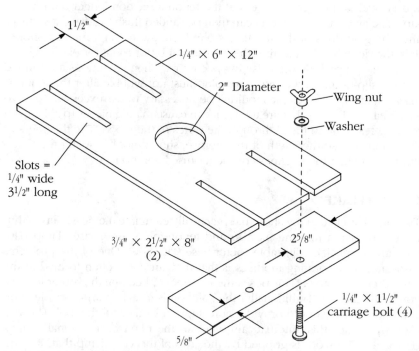

9-20 Construction details for the special router subbase used when mortising, forming dovetails, edge grooves, and similar chores.

tool can be positioned for mortise cuts that are off center as well as on center. The setup provides plenty of bearing surface for the router, and the guides ride snug against the jaws of the vise clamp to assure that the bit travels in a straight line. Cuts with a conventional router are started, as usual

9-21 Workpieces are gripped between the jaws of the vise clamp. With this setup, plenty of support area is available for the router.

9-22 Pictured is the assembled, special subbase. The self-adhesive measuring tapes are optional, but they do provide a quick means of setting the guides.

for this type of work, with the tool tilted and then slanted down so the bit can penetrate. With a plunge router, the bit is made to enter while the tool is firmly in position. Make repeat passes, when necessary, to deepen the cut.

CENTERING GUIDES

Centering guides, when they are designed as "part" of the router, assure that grooves or mortises will be centered in stock edges. The roller-type centering guides, shown in FIGS. 9-24 and 9-25, are attached directly to the router's base. Since it's not likely that there will be holes for this purpose, the process involves drilling two holes in the base of your router; either through holes, so the rollers can be attached with a nut and bolt, or tapped holes for just a screw. Either way, the holes must be equidistant from the router's centerline and they must be located so they can't cause a problem for the base itself.

The parallel rule design, detailed in FIG. 9-26, has its own mounting plate that substitutes for the router's subbase. Two advantages of this design

9-23 Once the guides on the subbase are set, you can be sure that all the cuts you make will be on the same centerline.

9-24 This roller-type centering guide has twin *wheels* that can be attached directly to the router's base or to a substitute base that you make.

are that once the bars of the jig have been adjusted to the thickness of the stock they can be locked in place, and the length of the bars add convenience when cuts are started from an end of the workpiece.

With these guides, the router is positioned by placing it on the work and then rotating it so the bearing edges of the guides rest against opposite surfaces of the work. Figures 9-27 and 9-28 show the parallel rule concept at work.

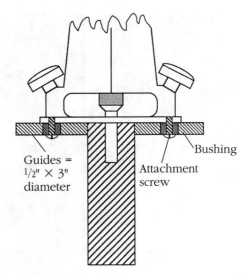

9-25 Mounting holes for the roller-type centering guides must be equidistant from the router's center. Ball-bearings can be substituted for the bushings.

Guides =
1/2" × 3"
diameter

Bushing

Attachment
screw

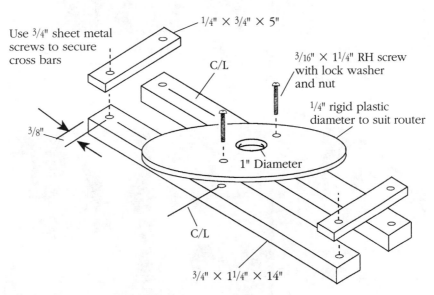

Use 3/4" sheet metal screws to secure cross bars

1/4" × 3/4" × 5"

C/L

3/16" × 1¼" RH screw with lock washer and nut

1/4" rigid plastic diameter to suit router

3/8"

1" Diameter

C/L

3/4" × 1¼" × 14"

9-26 Construction details for the parallel rule centering guides.

FORMING TENONS

Forming tenons is like making back-to-back rabbet cuts. Similar tenons are often needed on several of the project's components; for example, rails and stretchers for a chair. The tenons can be formed after the parts have been cut to width and length, but there is a better way. Select a piece of stock that is wide enough to accommodate the number of parts you need and cut it to correct length. If you need four parts, 2 inches wide, then the stock should be about 9 inches wide. The extra width is for the saw cuts that will be made to separate the stock into individual pieces.

9-27 Rotating the router will cause the arms of the parallel rule to bear against the sides of the workpiece, thus centering the router bit. The jig is easiest to use when mounted on a plunge router.

9-28 The long arms of the parallel rule guide add convenience and support when cuts must be started at the end of the workpiece.

Work with the largest straight bit you have. The depth of the cut, which is made on both sides of the work, should leave a projection of correct tenon thickness (FIG. 9-29). In this case, on 1½-inch material, the cut-depth is ½ inch, so the tenon thickness will be ½ inch. Be sure to mark the work correctly on both surfaces so the straightedge can be positioned accurately for the back-to-back cuts.

9-29 Pictured is the first step when similar tenons are required on multiple pieces. The shape is formed on material that is wide enough to be sawed into the number of parts you need. Notice the use of the outboard support block.

When the routing is complete, the stock is sawed into correct width components. If the tenon is for an open mortise and tenon joint (FIG. 9-30), then the job is done. If a true tenon with four shoulders is required, the individual pieces are clamped together or secured in a vise, and routing is done as if the assembly were a solid block (FIG. 9-31). Be sure that the pieces are in perfect alignment before using the router.

Mortises that are formed with a router bit will have round ends, so the tenon must be shaped accordingly (FIG. 9-32). The tenons can be used as is if you square the ends of the mortises with a chisel.

Mortising can also be done in miter cuts by using the same vise clamp and the special base setup (FIG. 9-33). The two mortises are then reinforced with a spline. Two factors are crucial in order for the joint to mate perfectly. The miter cut must be perfect and the cut end must be set perfectly flush with the jaws of the vise clamp. Figure 9-34 shows how the spline, in this

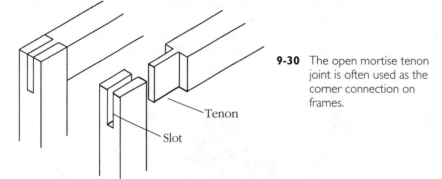

9-30 The open mortise tenon joint is often used as the corner connection on frames.

Tenon

Slot

9-31 To produce a true tenon, the individual pieces are clamped together and the final routing is done as if they were a solid block. Move the router slowly and be sure to keep it perfectly level.

case a strip of hardboard, is used in the joint. The spline should be sized for a slip-fit; its length should be a fraction less than the combined depth of the mortises.

Figure 9-35 shows another practical use for the vise clamp—securing a narrow strip for an edge cut. When necessary, the work can be elevated on blocks that are placed at each end of the vise clamp.

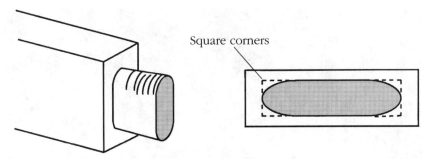

Square corners

9-32 Mortises made with a router bit have round ends so the tenon must be shaped to suit.

9-33 The vise clamp and the special subbase can also be used when you choose to form mortises in miter cuts.

A COMMERCIAL JIG

The jig shown in FIG. 9-36 was once known as "Morten the Jig," a cute name for a very practical idea. It's now a Porter Cable product, simply called a mortise and tenon template, but the changeover hasn't made it less useful. The jig is intriguing because it allows the user to create various types of mortise and tenon joints. Single, double, even triple mortise and tenon joints can be accomplished with impressive accuracy and relatively quickly, once you have become acquainted with the system (see FIGS. 9-37 and 9-38).

9-34 Like a tenon, the spline used in the miter joint should have rounded edges to suit the shape of the mortise.

9-35 Holding slim pieces of wood securely for edge shaping is another use for the vise clamp.

9-36 The frame assembly of the mortise and tenon jig has locating pins so the template can be located accurately for specific operations. Notice that the bit, which is supplied, has a ball-bearing guide located above the cutting edges.

9-37 The jig is secured to the workpiece, which is gripped in a vise. Here a double mortise is being completed. Allow the router to come to a full stop before lifting it from the work.

The heart of the accessory is the template and the way it can be registered on the frame assembly for particular cuts. Little, if any, measuring is required. Detailed instructions are provided with the jig package. Do not ignore the instructions. To get the most from a tool like this, it's best to go through a test period by trying procedures on scrap stock.

BECOMING A DOVETAIL EXPERT

Dovetails are hallmarks of quality construction. This eminence can be appreciated when you understand how superior a dovetail joint is when used

9-38 The router is moved so the guide on the bit starts to follow the template at the open end. Cutting must be done carefully to avoid damaging the first tenons that are formed.

as the connection between the front of a drawer and its side members. Each time you open a drawer the major stress on the assembly is where front and sides are joined. The pull on the drawer front is opposed by the other parts of the drawer and the weight of its contents. When you consider the number of times a kitchen drawer is opened during the life of the unit, you realize what the front-to-side connection must withstand.

The dovetail joint is an interlocking design. Pins, or tongues if you wish, have tapered sides that mesh with matching sockets. The connection opposes forces that tend to pull it apart, and it will do this even if the glue should fail.

Dovetail joints can be formed by hand with dovetail saws, coping saws, chisels, and such. They can also be formed by improvising procedures on a drill press, a table or radial arm saw, or even on a stationary shaper. The portable router, especially since the advent of modern techniques and jigs, allows anyone to quickly do professional work without sacrificing individuality and with minimum apprenticeship.

DADOES AND GROOVES BY DOVETAIL

Too often dovetail thinking is limited to case goods, drawers, and such, and the host of other possible applications are overlooked. A few of these are shown in FIGS. 9-39 through 9-42. Dovetail grooves, together with matching dovetail tongues, are excellent substitutes for regular dadoes and grooves when the strength required in the joint is crucial or when you have a yen to do something a bit different from the norm. The idea can also be used for sliding connections, in which case either the groove or the tongue must be "eased" enough to provide smooth movement. Areas where the sliding de-

9-39 Dovetailed grooves and tongues provide strong, interlocking connections.

9-40 Matching dovetail tongues can be glued in place for a permanent assembly or one of the components can be *eased* a bit for a sliding assembly. Sliding dovetails are often used on box projects like this one.

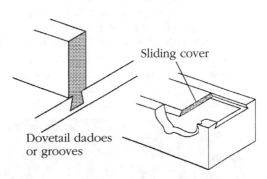

Sliding cover

Dovetail dadoes or grooves

9-41 This type of dovetail joint is often used in place of a mortise and tenon joint when, for example, making leg and rail assemblies.

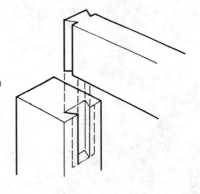

9-42 A similar application is often used to join legs to a pedestal. Dovetail cuts in round components can be formed accurately by using the fluting jig.

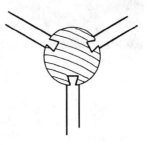

sign can be used include guidance systems for drawers, adjustable shelves or partitions, pullout shelves, sliding vertical supports that can be part of book or record racks, and projects like the box with the sliding cover that is shown in FIG. 9-40.

Any straight-line dovetail can be formed in line with or across the grain of workpieces by using a straightedge or one of the concepts demonstrated in chapter 6. When the cut is required in the edge of the workpiece, then the right setup is similar to what you would use when forming a mortise. The work would be sandwiched between extra pieces of wood to provide adequate support surface for the router; the router would be moved along with an edge guide.

When the length of the workpiece allows it, the vise clamp and special router base setup can be used. The dovetail groove is formed as shown (FIG. 9-43). The tongue needs a little more attention (FIG. 9-44). When creating the tongue, the work is placed between scrap pieces because cuts are required on both of its edges. Notice that only one of the adjustable base guides is used. When one side of the cut is finished, the router makes a second pass after it has been turned so the guide bears against the opposite edge of the jig. There is no need to readjust the guide.

9-43 Forming a dovetail in the edge of narrow work can be done accurately and with minimum fuss when the vise clamp and the special router subbase are used.

Figure 9-45 shows more details of forming dovetail grooves in edges. Notice that the same technique can be used to produce tongue and groove joints, an example of which is shown in FIG. 9-46.

READY-MADE DOVETAIL JIGS

There are quite a few commercial dovetail jigs available, and any router-user who doesn't own one or two is missing a good bet. All the fuss and

9-44 The dovetail tongue, sandwiched between scrap pieces, can be easily created using the vise clamp and the special router subbase. With this setup, the cuts are made without having to change the position of the guide on the subbase.

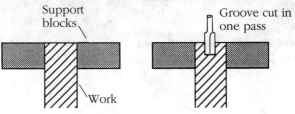

9-45 How dovetail grooves and tongues (or if you wish, slots and pins) are formed. With the work clamped between two support blocks, the tongue requires two passes; the router bit cuts into support blocks. This also applies to groove-type dovetails.

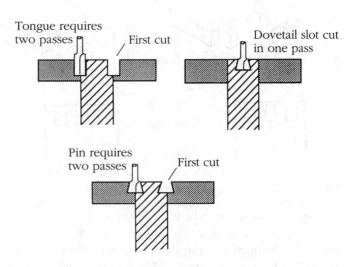

bother of layout and hand work, or the improvisations that are necessary when other tools are used, is eliminated when a dovetail jig is used to guide the router. The jigs allow the woodworker to produce the dovetail connections that are shown in FIG. 9-47, which are fairly standard designs in drawer

9-46 Sample of a tongue and groove joint that was formed following the same procedure for dovetailing.

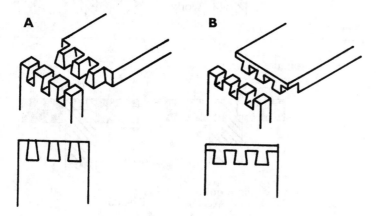

9-47 These are the most common dovetail joint designs. A is a through dovetail, B is a half-blind or French dovetail. Some jigs can be used only for the half-blind type; others allow you to do both.

and case constructions. The difference between them is more visual than anything else. The joint lines of the through dovetail are visible side and front. When the half-blind concept is used on a drawer, for example, the joint lines are visible only when the drawer is pulled out. "Seeing" the joint is not objectionable, since they are often deliberately exposed as a sign of dedicated craftsmanship.

Different names are used to identify parts of the dovetail joint, but those shown in FIG. 9-48 are generally acceptable. The configuration that is between sockets is called the *tail* and the projection itself is the *pin*. Confusion can be eliminated if you just think in terms of *pins* and *tails*.

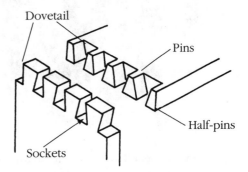

9-48 The common names for the parts of dovetails are labeled.

Dovetail

Pins

Half-pins

Sockets

When a jig like the one in FIG. 9-49 is correctly organized, you can produce as many precisely fitting dovetail joints as you like with each one taking only a few minutes to do. Production, however, is not the prime factor. For the woodworker who is usually involved with a single project, it's the ease with which dovetail expertise is acquired that counts. Being able to make joints like the one in FIG. 9-50 in "jig" time is pretty nice.

Using a jig isn't difficult, but you will be in trouble if you plunge right in without studying the manual supplied with the tool and without testing what you have learned on some scrap material. When you consider what the results can be, being an apprentice for an hour or two isn't asking much.

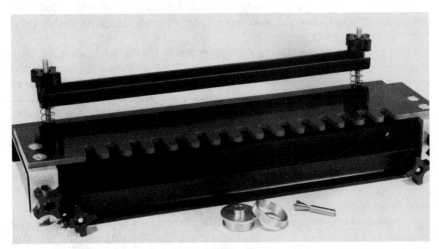

9-49 Typical jig that makes it easy to produce equally spaced, half-blind dovetails. It's best to mount the fixture on a sturdy board that, in turn, can be clamped to a workbench.

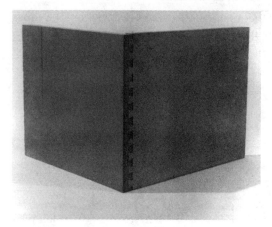

9-50 A precise dovetail joint that was formed quickly and accurately after thoroughly studying the instructions that came with the dovetail jig.

Important factors include the maximum width of stock the fixture can handle, whether you can work with one or several stock thicknesses, the size of the dovetails produced, and the size of the dovetail bit and template guide that should be used.

Most fixtures for half-blind dovetails are equipped with a finger template that is sized for producing ½-inch dovetails in stock thicknesses that might range from about ½ to 1 inch. It's often possible to purchase an additional template so the fixture can be used for smaller dovetails, usually ¼ inch. This makes it possible to work on thinner material, from about ⁵⁄₁₆ to ⅝ inch, but the work must be done with a template guide and a dovetail bit that are compatible with the finger template.

One of the intriguing features of this type of dovetail jig is that both components of the joint are clamped together in the fixture and the mating configurations (pins and tails) are cut at the same time (FIG. 9-51). When the shaping is complete, the two parts are ready for joining (FIG. 9-52). You can't go wrong if you follow the instructions for securing the parts in the jig so they have the correct relationship.

Start the cutting by placing the router firmly on the left end of the template but with the cutter clear of the work. Then turn on the machine and move it slowly into the work until the sleeve on the template guide makes contact with the finger template. Make the pass from left to right while allowing the template guide to control in and out movements. It's crucial to keep the router flat on the template throughout the pass. Turn the router off at the end of the cut, but keep it in working position until the bit stops turning. While keeping it on a horizontal plane, move it away from the template. For perfect results, you must avoid tilting the router at the start, during, and at the end of the cut. Tilting will damage the finger template as well as the workpieces.

Some manuals suggest that the cut be started by moving the router in a reasonably straight line along the outside edge of the fingers. The shaping is then finished by moving the router along the finger template in routine fashion. The thought is that the procedure will minimize the possibility that the part held vertically in the jig will be chipped as the router is moved in

9-51 A feature of the dovetail jigs is that the components are secured in correct relationship so both parts of the joint can be cut at the same time.

9-52 Can you imagine how long it would take to form these dovetail components by hand? And the care required to be as accurate? The only negative factor is that the jig decides spacing and size of the cuts.

and out of the template. This is not an unreasonable idea. At least it's one to consider should you encounter trouble in that area.

It's a good idea when using the jigs for drawer joints to identify the parts of the drawer and the inside and outside faces (FIG. 9-53). This will help you place the components in the jig so mating cuts will mesh and parts will appear as you want them to after assembly. The jigs can be used to produce drawers with flush or rabbeted (lipped) edges (FIG. 9-54). For the rabbeted design, the drawer front is prepared as shown in FIG. 9-55 before the dovetail shaping is done.

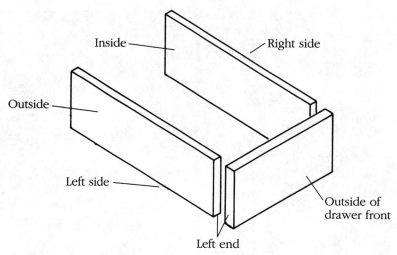

Inside — Right side

Outside —

Left side —

Outside of drawer front

Left end

9-53 Before doing dovetail work for a drawer, be sure to identify the parts and the faces that will be visible after the project is assembled so you'll know how to place the parts in the jig.

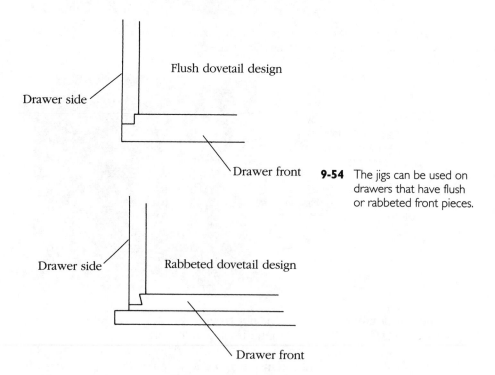

Flush dovetail design

Drawer side

Drawer front

9-54 The jigs can be used on drawers that have flush or rabbeted front pieces.

Drawer side — Rabbeted dovetail design

Drawer front

OTHER TYPES OF DOVETAIL JIGS

Dovetail jigs are fine tools that make it easy to produce dovetail joints of which you can be proud, however, they do have a drawback. They are designed primarily for half-blind dovetails, and the size and spacing of the cuts

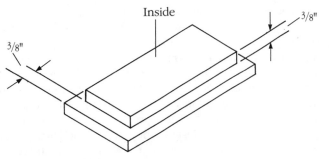

9-55 The stock for a rabbeted drawer front must be prepared a certain way. This is just an example. Check the owner's manual for particulars.

Inside

3/8"

3/8"

are dictated by the finger templates that can be used with the jig. To be more flexible and creative when dovetailing and to be able to decide the size, number, and spacing of the cuts (like those shown in FIG. 9-56), you must check out other dovetail jig concepts. These tools are more expensive, but they are structurally and conceptually impressive and produce advanced dovetailing work quickly. If you will only be using a dovetail joint occasionally, say in a drawer, you need not look here. However, if you want to do detailed dovetail joinery, then you ought to investigate the jigs mentioned on the following page.

9-56 To be completely flexible in the area of dovetail joinery, you can make cuts with hand tools or with a jig that allows you to size and space the dovetails as you choose.

The two new Leigh dovetail jigs, made in Canada but widely available in the United States, continue to interest and impress woodworkers. Improved models, like the 24-inch version (FIG. 9-57), and a 12-inch unit, are used in essentially the same manner as the "old" ones that were demonstrated in the first edition of *The Portable Router Book*. Both now have the same stock-thickness capacity (1¼ inch), but the basic difference between the two new units is the width of stock that can be gripped in the jig. The most intriguing aspect of the products is the flexibility that imposes little limitation on dovetail design. They can be used for conventional through or half-blind dovetails, but release the worker's creativity in terms of dovetail sizes, shapes, and, unlike fixed, one-piece finger templates, spacing (FIGS. 9-58 and 9-59).

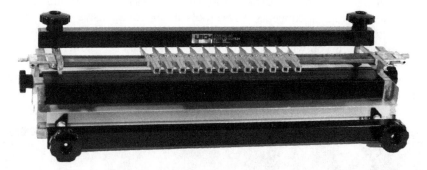

9-57 The new Leigh dovetail jigs, that come in 24-inch models (shown here) and a 12-inch one, have a 1¼-inch stock thickness capacity. They differ only in the work-width they can handle.

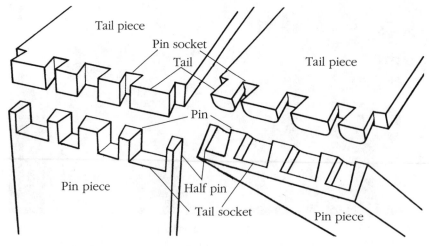

9-58 The basic dovetail joints that can be fashioned with the new Leigh dovetail jigs.

9-59 This is just an example of the flexibility you have when a jig like the Leigh product is available.

The guidance system of the Leigh jigs consists of individual fingers that have different configurations at each end and that are adjustable in relation to each other because they can be moved and then secured along a slide bar (FIG. 9-60). The fingers, after adjustment to suit your purpose, together with the slide bar form an assembly that can be flipped end-for-end or rotated depending on the current operation (FIG. 9-61). Once the finger assembly has been organized to your liking, cutting proceeds in routine fashion with a template guide installed in the router so the bit can follow the contours of the fingers (FIG. 9-62).

Forming through dovetails is a two step procedure. Shaping the tails is accomplished by guiding a dovetail bit along the straight end of the fingers (FIG. 9-63), while the pins are formed by using the opposite end of the fingers to guide a straight bit (FIG. 9-64). This may sound more complicated than it is. The operation actually goes pretty smoothly since no finger adjustment is needed when changing from one bit to the other.

Placing the finger assembly in correct position for each of the cuts is simplified because of the scales that are part of the jig design. You do have to be careful with bit projection when changing router bits. Many workers use two routers, one with the straight bit and the other with the dovetail bit, with each of them set to the correct depth of cut. This is a practical way to go especially when the same dovetail configurations are required on many components. Whether you work with one router or two, it's wise to make

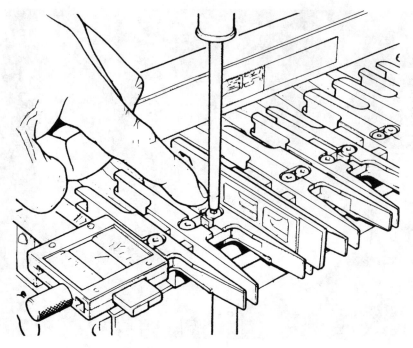

9-60 Individual fingers on the Leigh jigs can be positioned to suit operator's preference. They are secured with a special driver that is supplied with each jig.

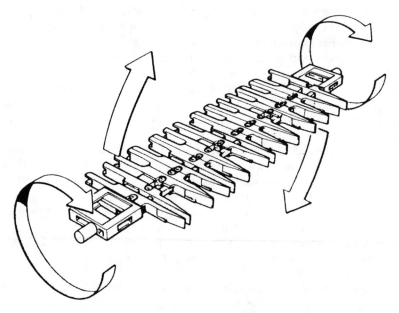

9-61 Fingers and slide bar form an assembly that can be flipped or rotated depending on what cut is involved. This is a major feature of the Leigh jigs.

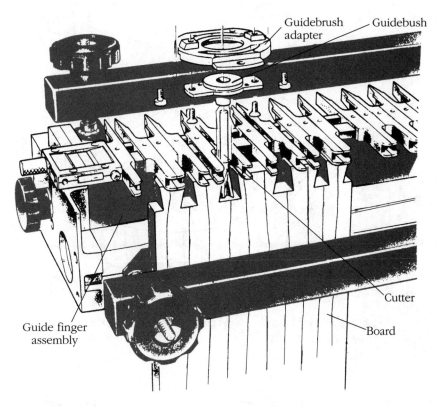

Guidebrush adapter

Guidebush

Cutter

Board

Guide finger assembly

9-62 Dovetailing proceeds in routine fashion after the finger assembly has been adjusted.

all of either the pin cuts or tail cuts (it doesn't matter which you start with), then the changeover to the cuts left to make.

Something new for the Leigh jigs is an optional accessory for producing variably spaced carcass type mortise-tenon joints (FIG. 9-65). The unit, which is mounted in place of the regular dovetail finger assembly, can be used for mortises and matching tenons up to 1½ inch × any length you choose. To use this new accessory, however, a plunge router is needed with a two-flute, up-cut spiral bit installed.

The instructional materials that are provided with the Leigh jigs are examples of how owners' manuals should be written. They are clear, very detailed, include a troubleshooting chart, and provide instructions that go beyond basic dovetailing, such as providing added information on angled dovetails, cogged dovetails, end-to-end connections, using the jigs for template work, and more.

Another example of an advanced dovetail jig from the Woodmachine Company, is shown in FIG. 9-66. This tool is a husky concept with a cast-iron base, clamps and handles of steel, and templates of ¼-inch-thick aluminum (FIG. 9-67). The 65-pound machine can handle stock up to 1 inch thick and up to 16 inches wide. The dovetail bit, which is supplied with the tool, is guided by a ball-bearing. This eliminates the need for a template guide. The

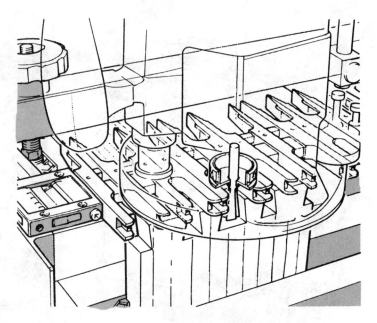

9-63 The tails are fashioned with a dovetail bit that is led along the straight ends of the fingers by a template guide. The router (phantom lines) must always be flat on the fingers.

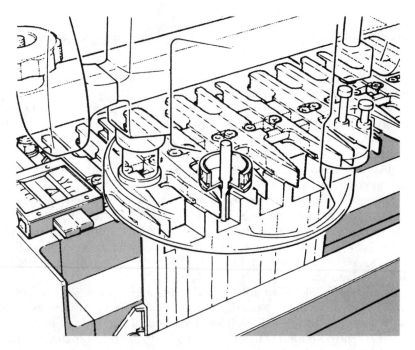

9-64 The pins are routed with a straight bit after the finger assembly has been flipped. No adjustment of the fingers is necessary.

9-65 Optional accessory for the new Leigh jigs is a finger assembly that is used to produce variably-spaced, carcass-type mortise and tenon joints. The unit is used with a plunge router.

9-66 Actual cutting with the Woodmachine Company's dovetail jig doesn't differ from other tools in the dovetail category except that the dovetail bit is guided by a ball-bearing that follows the fingers on the template. However, it isn't necessary to have a template guide in the router's subbase.

9-67 The basic template for the Woodmachine Company's dovetail jig is used for standard ½-inch dovetails. Other templates, all of ¼-inch-thick aluminum, are available.

Other types of dovetail jigs   **157**

basic template is for producing ½-inch machine-type dovetails, but optional templates are available for flexibility in dovetail design (FIGS. 9-68 to 9-70). A special template allows the production of precise finger-lap joints, like those shown in FIG. 9-71.

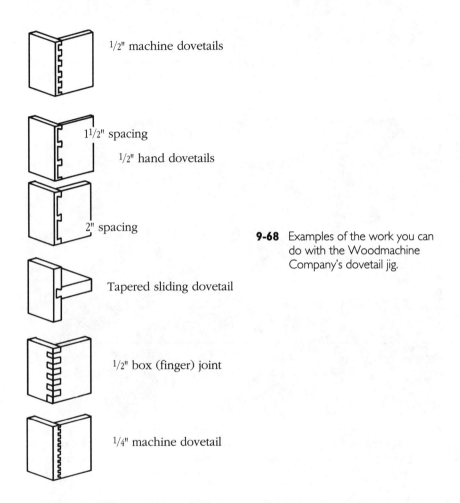

1/2" machine dovetails

1¹/2" spacing

1/2" hand dovetails

2" spacing

Tapered sliding dovetail

1/2" box (finger) joint

1/4" machine dovetail

9-68 Examples of the work you can do with the Woodmachine Company's dovetail jig.

Like all tools of this nature, it is crucial that the user study and diligently follow the operational information to get optimum results. When errors occur, it's usually because of the desire to start work on the project before thorough understanding of the tool.

If you are interested in an easy-to-learn method for producing through dovetails, one that allows some degree of flexibility, look into the Keller Dovetail System. There are several models available, each of them consisting of two ½-inch-thick, ruggedly constructed, precisely machined aluminum templates, plus mounting hardware and dovetail bits with shank-mounted pilot bearings so the router doesn't have to be equipped with a template guide (see FIGS. 9-72 through 9-74).

9-69 Standard half-blind dovetails fit as if they were cut with a laser. The machine can cut stock up to 16 inches wide, which is capacity enough even for most case goods.

9-70 A feature of the Woodmachine Company's dovetail jig, like others of its type, is that you can be creative with size and spacing of cuts.

The hardware is used to secure each of the templates to a 1½-inch mounting block whose width and length depend on the templates being used. This must be done accurately, so to assist with the process a scribed line is provided on the bottom surface of the templates, parallel to the long edges (FIG. 9-75). The mounting holes for the pin template are slotted so some to or fro adjustment is feasible if required to get tails and pins to mate correctly.

Cutting operations are straightforward. It's a good idea to first form the tails using the dovetail bit. Then use that component, as shown in FIG. 9-76, to mark one or two tail locations so the second component, the pin board, can be situated accurately with the pin template. When making multiple parts, situate a stop block on the template or mounting board so the

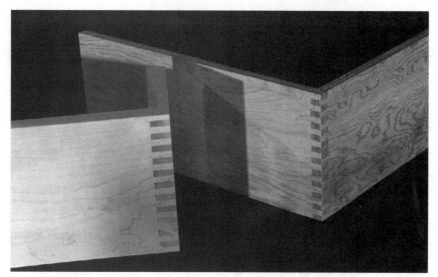

9-71 With a special template, the Woodmachine Company's dovetail jig can be used to produce finger lap joints. Like the dovetail, this joint design is often left exposed. The interlocking fingers and the large glue area make this a very strong connection.

9-72 Keller dovetail templates are husky, ½-inch-thick, precisely machined aluminum. One template, shown here, is used with a dovetail bit.

9-73 The mating Keller template is used with a straight bit to form the pins.

9-74 Both of the jigs supplied with the Keller templates have shank-mounted pilot bearing so they can operate without the need of a template guide.

Mounting block

Mounting screws (4)

Dovetail template

Scribed line
(underside of
template)

9-75 Each of the Keller templates is mounted on a 1½-inch-thick block. The work is secured in a vise; block-mounted templates are clamped to the work.

Slot for mounting screws
allows adjustment

Pin template

Scribed
line

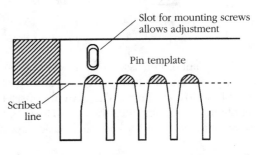

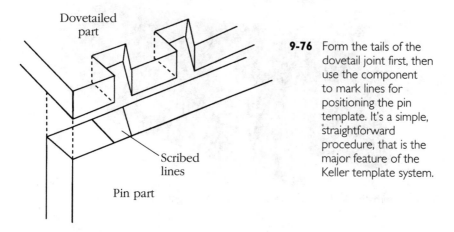

Dovetailed part

Pin part

Scribed lines

9-76 Form the tails of the dovetail joint first, then use the component to mark lines for positioning the pin template. It's a simple, straightforward procedure, that is the major feature of the Keller template system.

initial marking procedure doesn't have to be repeated. The length of the templates does not limit the width of work that can be dovetailed. When work-width exceeds template-length, just reposition the template and continue routing.

The templates can be used for rabbeted pins, but only if the mounting block is modified to receive a pre-rabbeted workpiece as shown in FIG. 9-77. The Keller Company now offers three sizes of templates so you can choose a set that is most appropriate for your needs. Table 9-1 lists what is available and suggests collet and horsepower requirements.

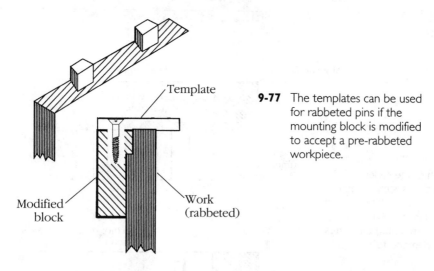

Template

Modified block

Work (rabbeted)

9-77 The templates can be used for rabbeted pins if the mounting block is modified to accept a pre-rabbeted workpiece.

The JointMaster, issued by Leichtung, is a practical and flexible joint-making concept that is used to form, among other things, mortise-tenon and half-blind dovetail joints. The system consists of a mounting plate and the set of integrated templates shown in FIG. 9-78, plus a router base kit and necessary template guides and bits. Basic capacity is boards up to 7 inches

Table 9-1 Keller dovetail templates

Model	Template length	Router collet	Max. stock thickness		Router hp
			Plain pins	Rabbeted pins	
1601	16"	¼"	¾"	1"	¾
2401	24"	⅜"–½"	1"	1½"	1¼
3600	36"	½"	1"	1½"	1½

9-78 The JointMaster system consists of an ingenious assortment of templates that are used to form a variety of dovetail joints plus mortise-tenon connections.

wide, but going beyond this is just a matter of repositioning the templates and continuing to rout.

Before you can put the unit to use, you must provide a sturdy, bracket-type stand (detailed in FIG. 9-79) that can be gripped in a vise or clamped to a bench. The mounting plate is secured to the stand with supplied hardware; templates for particular joint formations are added to the mounting plate (FIGS. 9-80 and 9-81).

As with most products of this nature, it's important to obey the instructions in the owner's manual and to assure accuracy of the finished product by making a test cut in scrap stock whose thickness equals that of the project material. The ¼-inch straight bit that is supplied is used to cut mortises and tenons in single or multiple fashion. You can also produce half-blind or through finger joints but this requires an additional ⁷⁄₁₆-inch-diameter straight bit. It's recommended that the extra bit be purchased from the supplier since it is manufactured to tolerances that are exactly right for the JointMaster. All-in-all, the JointMaster system lends itself to good performance. It takes a little time to get acclimated to the procedures, but once you've learned them, producing multiple, identical joint forms is no problem.

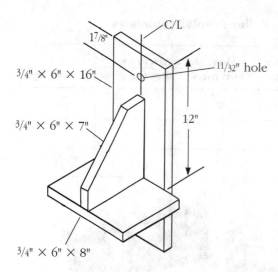

17/8"

C/L

3/4" × 6" × 16"

11/32" hole

3/4" × 6" × 7"

12"

3/4" × 6" × 8"

9-79 The JointMaster system requires the buyer to supply the mounting stand, whose construction details are shown here. It can be made of solid wood or a good grade of plywood. Assemble the project with glue and #10 × 1¾-inch flathead screws.

9-80 The main mounting plate of the JointMaster system is secured to the stand; templates are then added to the plate. Here the template for half-blind dovetails is being used.

9-81 The JointMaster system's tenon template has centerlines on it so you can align the lines on the template with the centerlines on the tenons. The system calls for careful study of the detailed manual that is supplied with the product.

Plate, or *biscuit* joinery, has been around for some time in industry but because of new, small, *plate Joiner* power tools, the technique has now become common in small woodworking shops too. The Craftsman *Bis-Kit* accessory has now made the novel but highly practical joint-forming system of biscuit joinery available to portable router users. The kit consists of a heavy carriage that accommodates the router and a carbide-tipped slot cutter that is installed like any router bit.

The joiner is used by keeping the carriage face in contact with a secured workpiece and then sliding the router forward so the cutter forms the half-ellipse cut that receives the biscuit (FIG. 9-82). The carriage allows adjustment of the width and depth of the cut so any of the standard biscuits that are available can be used (see Table 9-2).

While biscuit joints are popular for connecting parts edge-to-edge, there is enough flexibility in the system so the technique can be used in any of the common joints shown in FIG. 9-83. Making layouts for accurate cutting is pretty simple. Simply butt the mating parts together, then draw a common centerline across them. Position the tool so its registration mark lines up with the layout line before moving the router forward to make the cut.

DROP-LEAF TABLE JOINT

A table with hinged leaves requires little floor space when not in use, but can be "spread" to accommodate quite a few people. Often the edges on the table and leaf are left square and butt hinges or a continuous (piano) hinge are used as the pivot device. However, the true dropleaf joint or *rule joint* that is found on many pieces of modern and traditional furniture is the neater, more professional approach.

As you can see in FIG. 9-84, the secret of the joint is the way the quarter cuts complement each other. When the cuts are made correctly, the drop

9-82 Craftsman's Bis-Kit accessory allows a router to be used for "biscuit" joinery. Check for adaptability before buying since the unit may not be usable on the router you own.

Table 9-2   Biscuit sizes

#	Thickness	Width	Length
"0"	⁵⁄₃₂"	⅝"	1¾"–1¹³⁄₁₆"
"10"	⁵⁄₃₂"	¾"–¹³⁄₁₆"	2⅛"–2⅛"
"20"	⁵⁄₃₂"	¹⁵⁄₁₆"–1"	2⁵⁄₁₆"–2⅜"

(Choose largest size the joint will accommodate)

leaf swings up easily and extends the table surface on a smooth line. The edge of the drop leaf is shaped with a cove bit, while the table's edge is shaped with a rounding-over bit. The bits must have the same radius, and the depth of the cut on each component must be exact.

The hinge that is commonly used is called a *back flap* and is installed so the knuckle fits a shallow groove that is cut into the wood. You can form the groove with a core box bit or form a suitable mortise for it by using a small chisel. A similar hinge that has one leaf slightly bent is available and can be installed without a groove for the knuckle.

Because of the close tolerances required for optimum results, it's a good idea to test the router setups, even the hinge mounting, on some scrap

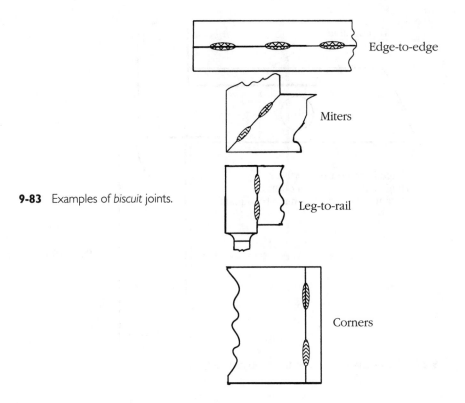

Edge-to-edge

Miters

9-83 Examples of *biscuit* joints.

Leg-to-rail

Corners

pieces before working on the project material. Remember that there must be some clearance between the shaped edges so that they won't rub against each other. A trick that provides adequate clearance is to place a length of wrapping paper between the edges of the table and the leaf when you are assembling them.

SPLINES

Splines are strips of material that are used to strengthen various joints (FIG. 9-85), but they can do more than just add strength. The use of splines helps to hold parts in alignment during assembly procedures. They are often made of a contrasting material and left exposed as a decorative detail. With a router, the best way to form the grooves for splines is to work with a slotting cutter. Because blades for the cutter are available in several thicknesses, there is some liberty in deciding how thick the spline will be.

Plywood, because it has strength in all directions, and hardboard, which is grainless, are good materials to use for splines. When you use lumber to make special splines, be sure the grain of the wood runs across the spline because wood splits more easily with the grain than across the grain. Splines should slip-fit into the grooves. Having to force them into place will only complicate assembly procedures and can even create stresses that might cause components to split.

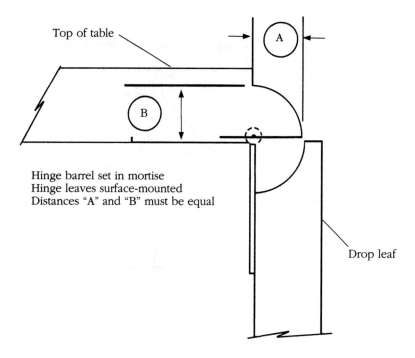

Top of table

A

B

Hinge barrel set in mortise
Hinge leaves surface-mounted
Distances "A" and "B" must be equal

Drop leaf

9-84 Construction details on the drop leaf table joint.

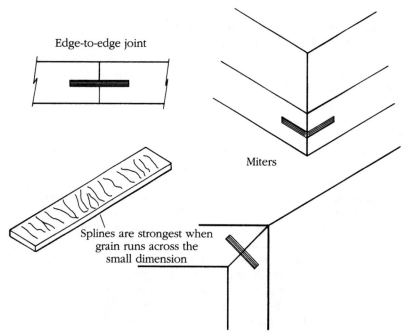

Edge-to-edge joint

Miters

Splines are strongest when
grain runs across the
small dimension

9-85 Splines can be used to reinforce many types of joints.

168 The router as a joinery tool

Another type of spline, actually an inletted piece, is shown on the miter joint in FIG. 9-86. This can be installed after the parts have been glued together. The recess for it is really a blind dado that runs at right angles to the joint line. Make the cut by moving the router along a straightedge that is clamped at a suitable point across the joint pieces. Stop blocks are not needed because the spline, or inset piece, will be formed to suit the cut. It's okay to work to lines that you mark on the work.

The depth of the recess should be a bit less than the thickness of the spline material. Sand the spline after the glue has had time to set so it will be flush with adjacent surfaces.

9-86 This type of spline is actually a contrasting piece glued into a blind dado that is cut at right angles across the joint. Cut the spline to suit after the recess has been formed.

10

Special jigs you can make

Like many power tools, the extent to which a portable router can be used depends on the viewpoint of the user. The tool can be limited to some basic functions or it can be organized in various ways to serve in capacities even the manufacturer hasn't envisioned. This is where a jig, or fixture if you wish, comes in. The homemade jig might serve as an accessory that is not available commercially or it might duplicate to some extent an existing, costly product or one whose frequency of use would not justify its purchase.

To a degree, all homemade jigs are improvisations. This doesn't mean you should be haphazardous when making them. Even if you invent a jig that solves a one-time problem, you must be sure that the tool/jig combination—in concept and in construction—is as safe to use as any standard piece of equipment. Any jig you make should become part of your router workshop. This is easy to accept if the jig will see frequent use, but even the one-time problem solver may be needed again.

FLUTING JIG

The fluting jig shown in FIG. 10-1 is an adjustable holding device for spindle-type or square workpieces on which you can make longitudinal cuts. The jig has a movable *tailstock* and centers at each end that adjust vertically so the unit can accommodate workpieces of various lengths and diameters. The sides of the jig provide support for the router.

Figure 10-2 shows how the jig is used. The work is secured between the centers. The router, fitted with a cutter of your choice, is centered over the work and guided through the cut by an edge guide. The special subbase that was shown in Chapter 9 can also be used with the fluting jig. The cut can be of equal depth throughout its length or, because it's

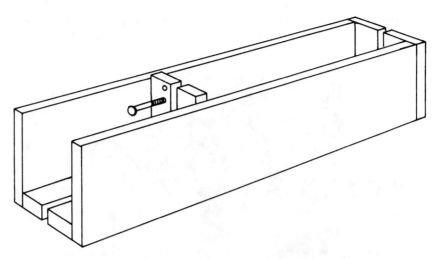

10-1 The fluting jig is basically a trough that, like a lathe, has a fixed headstock at one end and an adjustable tailstock at the other end, making it possible to accommodate workpieces of various lengths.

10-2 The tailstock is situated to secure the work between the centers of the jig. The router is placed so the centerline of the bit is over the longitudinal centerline of the work. Cuts can be uniform in depth or tapered, depending on how you decide the vertical position of the centers.

possible to set one center a bit lower than the other, it can be tapered. To make stopped cuts, you just tack-nail a strip of wood across the sides of the jig (FIG. 10-3).

10-3 The cuts can be across the full length of the work, or they can be "stopped" by tack-nailing a strip of wood across the edges of the jig.

Construction details for the fluting jig are shown in FIG. 10-4. The length of the jig will easily accommodate legs for chairs and coffee tables, but its length can be extended to work on even longer components. The centers are made by grinding points on the end of ¼-inch × 2-inch bolts. The holes in the vertical member of the tailstock are for an indexing pin that keeps the work in a fixed position when routing is done. The indexing pin, which is tapped into the end of the workpiece that is positioned in the jig, is made by sharpening the point on a 16d nail (FIG. 10-5).

For spacing cuts on a workpiece, use the system shown in FIG. 10-6 to mark your cuts. This system simply employs a strip of paper wrapped around the work and cut so its length equals the circumference of the piece. To determine spacing of the cuts, you fold the strip as needed. For example, for cuts that are 180 degrees apart, you fold the paper once; for cuts that are 45-degree spacing apart, you fold the paper twice; and so on. Once the cuts are marked on the strip of paper, put the strip back on the workpiece and mark the fold lines on the work. Position the work in the jig by lining up a mark with the center of the slot that is in the tailstock. Then tap in the indexing nail to hold the work still.

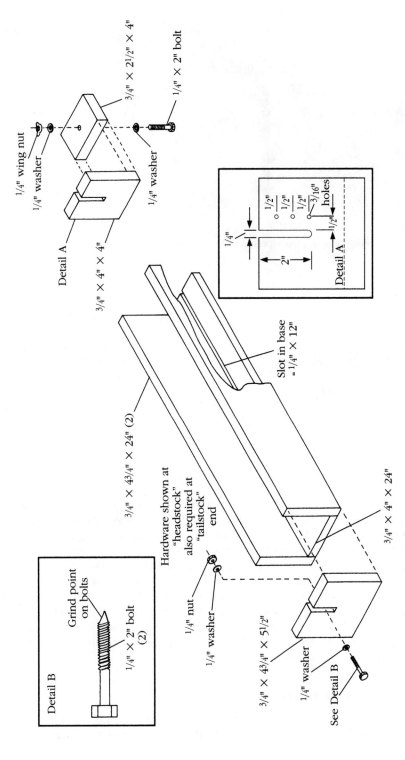

1/4" wing nut

1/4" washer

$3/4" \times 2^{1}/_{2}" \times 4"$

1/4" × 2" bolt

1/4" washer

Detail A

$3/4" \times 4" \times 4"$

1/4"

1/2"
1/2"
1/2"

3/16" holes

2"

1/2"

Detail A

Slot in base = $1/4" \times 12"$

$3/4" \times 4^{3}/_{4}" \times 24"$ (2)

Hardware shown at "headstock" also required at "tailstock" end

1/4" nut

1/4" washer

$3/4" \times 4" \times 24"$

Detail B

Grind point on bolts

$1/4" \times 2"$ bolt (2)

$3/4" \times 4^{3}/_{4}" \times 5^{1}/_{2}"$

1/4" washer

See Detail B

10-4 Construction details for the fluting jig. The trough of the jig can be made longer if you wish.

10-5 After the work is mounted, the index pin, which is a sharpened 16d nail, is tapped into place so the work will not rotate as the cut is made.

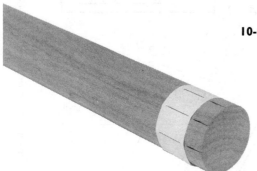

10-6 A strip of paper whose length matches the circumference of the workpiece is used to mark the spacing between the cuts. The marks on the work are lined up with the center of the slot that is in the jig's head-stock or tailstock.

Results depend on the cutter you use. Core box bits, V-cutters, straight bits, and such can be used for simple fluting; pilotless bits for decorative grooves; and so on. If you work with a mortising bit, you can form flats on cylinders. Even jobs like forming stopped dovetails in pedestals for leg attachment can be done accurately by using the fluting jig (FIG. 10-7).

The jig is not limited to working on cylinders. Square stock can also be mounted for fluting or shaping in various ways. The flutes in the example shown in FIG. 10-8 were formed with the router positioned so the bit would cut away from the centerline of the workpiece. Work with a large diameter straight bit or a mortising bit and you can "cut corners" to change the shape of a square into an octagon.

JIGS FOR TAPER CUTS

Legs for chairs, tables, and some other projects are often cut so they taper, either partially or for the full length of the workpiece. A tapering jig is re-

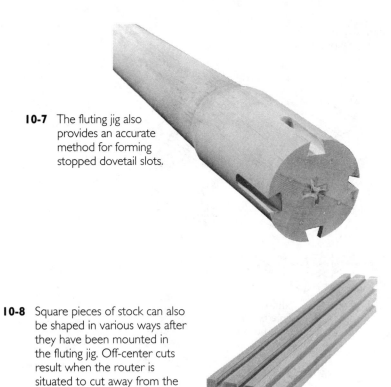

10-7 The fluting jig also provides an accurate method for forming stopped dovetail slots.

10-8 Square pieces of stock can also be shaped in various ways after they have been mounted in the fluting jig. Off-center cuts result when the router is situated to cut away from the centerline of the workpiece.

quired when this kind of work is done on a sawing machine. This also holds true when a router is used. For a router, the jig is assembled as a trough that suits the size of the work at hand. The work is placed in the trough and elevated at one end with a height block or, preferably, a wedge and secured at the opposite end with a clamp (FIG. 10-9). Mark the lines of the taper beforehand so the work can be elevated with the wedge at one end to suit the cut that is required. You are making a level cut, but it results in a taper simply because the work is secured in a tilted position.

Wedges are suggested over height blocks because they can be adjusted in terms of how high the tapered end of the work must be in the jig. Extra wedges are wise because they can be used between the sides of the workpiece and the jig to provide rigidity as the work progresses. The same jig can be used regardless of whether the workpiece will be tapered on one, two, or all four sides.

Another type of taper jig for decorative surface cuts can be organized in a radial pattern (FIG. 10-10). The jig isn't more than a ramp that allows the bit to cut deeper at one end of the pass than at the other. Minimum and maximum cut depths are controlled by the slope of the ramp and the projection of the bit. Jigs like this should be made so the router can be used with a template guide; therefore, the width of the guide slot in the jig should suit

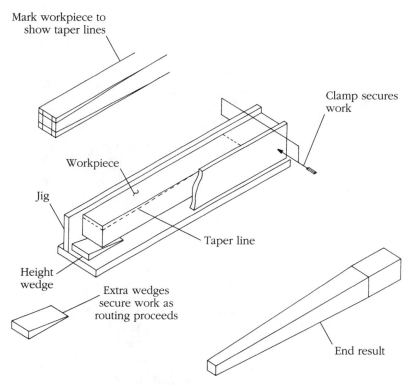

Mark workpiece to show taper lines

Clamp secures work

Workpiece

Jig

Taper line

Height wedge

Extra wedges secure work as routing proceeds

End result

10-9 This taper jig is a trough-type design that is sized to suit the work at hand. The router rides on the top edges of the side pieces. The path of the cutter is horizontal, but the workpiece is tapered because its height at one end of the jig is increased with a height block or wedges.

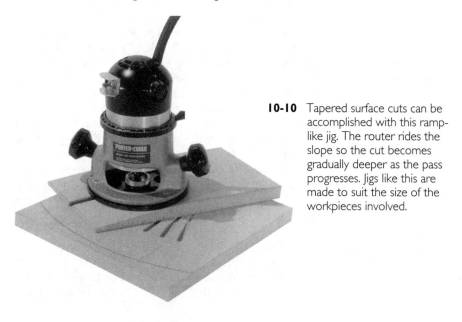

10-10 Tapered surface cuts can be accomplished with this ramp-like jig. The router rides the slope so the cut becomes gradually deeper as the pass progresses. Jigs like this are made to suit the size of the workpieces involved.

the outside diameter of the sleeve on the template guide (FIG. 10-11). By drilling a series of holes on the centerline of the slot, the one jig can be used on various size workpieces.

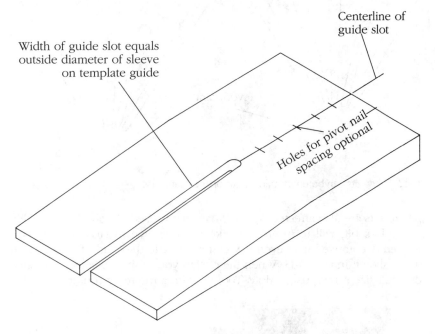

Width of guide slot equals outside diameter of sleeve on template guide

Centerline of guide slot

Holes for pivot nail spacing optional

10-11 The jig can be made variable for different work-sizes by drilling a series of holes on the centerline of the guide slot. The thickness of the ramp at its small end can't be less than the length of the sleeve on the template guide.

To set up for the operation, first draw a circle on the work and then mark the circumference, or the arc as the case may be, so the slot in the jig has points of reference. The pivot point of the jig (a nail) is at the center of the same circle (FIG. 10-12). The jig is set for each cut by aligning the center of the slot with the mark on the circle. The cut is stopped when the forward edge of the router's subbase meets the marked circle. The cuts do not have to be the same length. Variations are possible simply by marking different stop points on either the work or the jig (FIG. 10-13).

10-12 The jig is situated for each cut by lining up the center of the guide slot with location points marked on the circumference of a circle. The pivot point (arrow) is just a small nail.

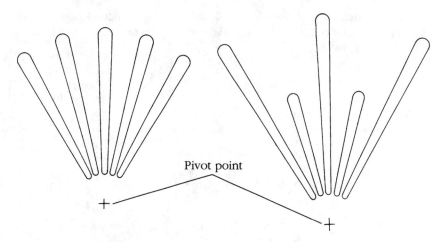

10-13 The cuts can be uniform in length or they can vary.

Pivot point

Results are also affected by the cutter that is used. A round end bit, like a core box bit, will form grooves like those shown in FIG. 10-14, while a pointed decorative bit or even a V-cutter will add the kind of center detail that is shown in FIG. 10-15. When the design you wish to create is too small for a full-size router, use a Moto-Tool or even a trimming router.

10-14 The bit you use affects the cuts you get. These tapered grooves were formed with a small core box bit.

A PERIPHERAL ROUTING JIG

Routing into the edges of circular pieces and others with regular or irregular curves can be done easily and accurately if you make a peripheral routing jig like the one shown in FIG. 10-16, and demonstrated in FIG. 10-17. The jig, which secures a router or laminate trimmer motor, is clamped to a workbench. The work is elevated on a block so it will be at the correct height and moved past the cutter while being guided by the post in the base of the jig. The post also assures a uniform depth of cut.

Construction details for a typical peripheral jig are shown in FIG. 10-18. It's best to make the holding block from a solid piece of stock. The tool hole must suit the diameter of the router motor and can be formed in various

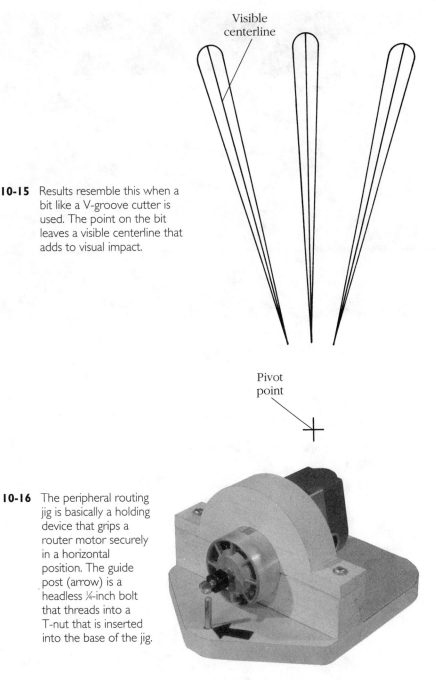

Visible
centerline

10-15 Results resemble this when a bit like a V-groove cutter is used. The point on the bit leaves a visible centerline that adds to visual impact.

Pivot point

10-16 The peripheral routing jig is basically a holding device that grips a router motor securely in a horizontal position. The guide post (arrow) is a headless ¼-inch bolt that threads into a T-nut that is inserted into the base of the jig.

ways. Make a template and use a router equipped with a template guide and straight bit to do the cutting. If you have the equipment, the hole can be formed on a drill press with a hole saw or fly cutter. It can also be formed by hand with a coping saw. After the block is formed, drill pilot

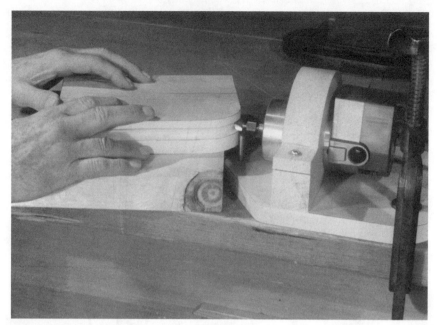

10-17 This figure demonstrates a fairly typical application for the peripheral jig. The work is moved past the secured jig.

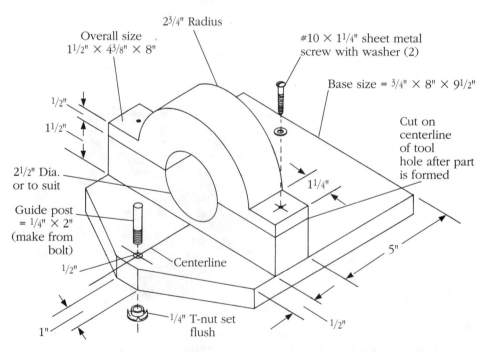

23/4" Radius

Overall size
11/2" × 43/8" × 8"

#10 × 11/4" sheet metal
screw with washer (2)

Base size = 3/4" × 8" × 91/2"

1/2"

11/2"

Cut on
centerline
of tool
hole after part
is formed

21/2" Dia.
or to suit

11/4"

Guide post
= 1/4" × 2"
(make from
bolt)

5"

1/2"

Centerline

1/2"

1/4" T-nut set
flush

1"

10-18 Construction details for a typical peripheral routing jig. The tool hole must be sized to suit the diameter of the router's motor.

holes for the sheet metal screws and then saw the piece on the centerline of the tool hole. The saw kerf should be about ⅛ inch wide so that when the screws are tightened, the upper part of the holder bears down on the motor to hold it securely.

Figure 10-19 shows decorative surface grooving being done on a curved component. It's important to keep the work flat on the height block and firmly against the guide post throughout the pass. Depth of cut can be achieved by positioning the motor in the tool holder, not by excessive projection of the bit.

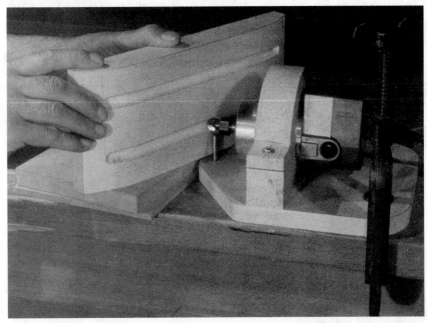

10-19 Another example of how the peripheral jig can be used. Notice that the height block or blocks under the work must raise it enough so its bottom edge clears the top surface of the jig's base.

The operation can also be reversed (FIG. 10-20). The work can be held securely with a clamp while the jig is moved to make the cut. It's important to work on a clean, flat surface so the jig can move smoothly.

THE SWIVEL JIG

The swivel jig (FIG. 10-21) consists of a sturdy, raised platform that supports a rotatable indexing plate with an attached toolholder that is mounted on pivots so the tool can be swung in an arc (FIG. 10-22). Because the tool can be rotated as well as swiveled, decorative surface cuts are easy to produce (FIG. 10-23).

The jig must be made very carefully for cuts to be precise and have correct radial alignment. Notice, in the construction details shown in FIG. 10-24,

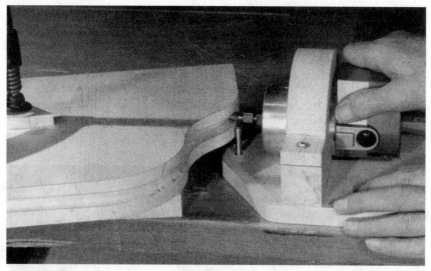

10-20 The peripheral jig can also be used by clamping the work and moving the jig to make the cut. The guide post acts something like a pilot to control the depth of the cut.

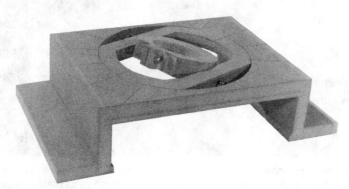

10-21 The swivel jig consists of a substantial base that supports an indexing plate and a pivot-mounted clamp for a router's motor.

that the top of the platform is a two-piece assembly. The diameter of the circle in the top part is 9 inches, while the bottom piece has an 8-inch circle. When the parts are assembled, the difference in the diameters supplies a ½-inch ledge for the indexing plate to rest on.

Make the indexing plate with care so the only movement it can have in the platform is a circular one. To reduce the bulk of the jig, the toolholder is sized to suit the motor of a laminate trimmer rather than the motor of a full-size router. A jig of this type can also be made to suit a Moto-Tool or a high-speed grinder.

When laying out the index marks, be sure that the first ones are exactly on the perpendicular diameters of the platform and the indexing plate. Other marks can be located with a protractor. Accurate marks are crucial

10-22 The motor can be swung in an arc and rotated to cut on various radii. If you make a mistake, and the index plate has any play, use a small-C-clamp to keep it in position for each of the cuts you make.

10-23 Just a few examples of the decorative surface cuts that can be made by working with a swivel jig.

because they are used to control the radial alignment of the cuts. Also, be sure the toolholder is centered perfectly inside the index plate.

Typical cuts that can be produced with the swivel jig are shown in FIGS. 10-25 and 10-26. Results depend on the number of cuts, their spacing, the depth of the cuts, and the router bit that is used. Experiment with a bit like a V-groove cutter before researching effects that can be achieved with straight bits, core box units, and pilotless decorative bits.

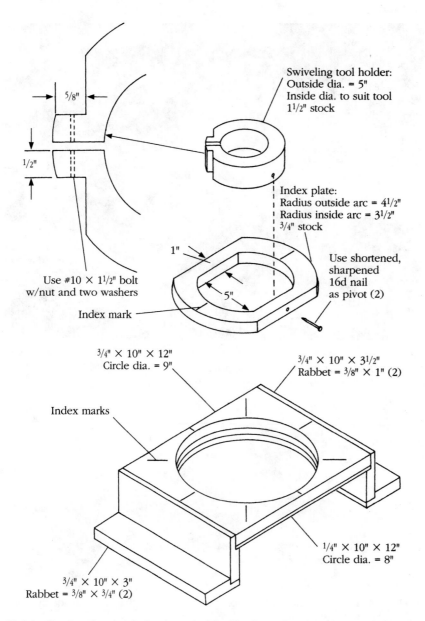

5/8"

1/2"

Swiveling tool holder:
Outside dia. = 5"
Inside dia. to suit tool
1¹/2" stock

Index plate:
Radius outside arc = 4¹/2"
Radius inside arc = 3¹/2"
³/4" stock

1"

Use shortened,
sharpened
16d nail
as pivot (2)

Use #10 × 1¹/2" bolt
w/nut and two washers

5"

Index mark

³/4" × 10" × 12"
Circle dia. = 9"

³/4" × 10" × 3¹/2"
Rabbet = ³/8" × 1" (2)

Index marks

¹/4" × 10" × 12"
Circle dia. = 8"

³/4" × 10" × 3"
Rabbet = ³/8" × ³/4" (2)

10-24 Construction details for the swivel jig. The jig performs perfectly only if all the parts are made to mate correctly.

AN OVERARM FOR THE PORTABLE ROUTER

There are times when using a portable power tool when it would be nice to have an extra hand. This thought applies as equally to a hand-held router as to other machines. Even though a major feature of the machine is its maneuverability, there are times when it would be nice for the little powerhouse

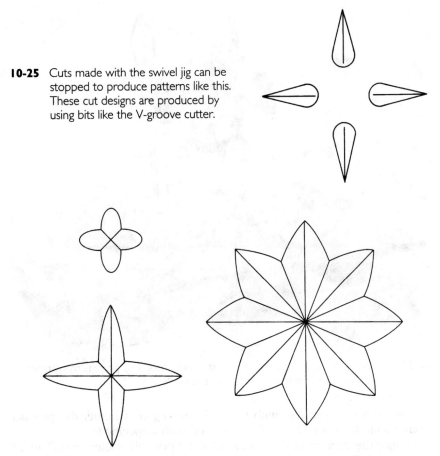

10-25 Cuts made with the swivel jig can be stopped to produce patterns like this. These cut designs are produced by using bits like the V-groove cutter.

10-26 Elaborate designs like this are possible when you use the swivel jig.

to sit still while hands are free to guide the work. Accuracy is important since the scope of the tool ranges from forming woodworking joints to producing intricate silhouettes. My solution was to build a jig so the router could be used as a stationary machine—in essence, a versatile overarm pin router (FIG. 10-27).

There are many accessories for portable routers, but most add or simplify a specific operation. The overarm pin router is able to handle the bulk of operations normally done by hand-holding a router but without the assortment of work holders, jigs, and fixtures that are otherwise required. It adds convenience, decreases setup time, and contributes to accuracy, since the chore of simultaneously controlling the tool and maneuvering it is eliminated.

An industrial overarm router, which usually has an integral power head (and a price range starting at about $2,000), is pretty standard in furniture factories and top-notch cabinet shops. Lately, however, units have been developed with the amateur woodworker in mind. With these new additions,

10-27 The overarm sets up the router as a stationary tool. This permits many practical chores but with hands free to control the work.

prices are not so startling, with much depending on whether the product comes with its own router or must be used with a specific tool.

Once the concept of the overarm for the portable router crystallized, I experimented a bit with the means of mounting the router. The first design used a split clamp to grip the router without its base. However, every time I needed to change the cut depth, I had to loosen the clamp or move the support arm. The solution was to install the router on its base on a plate that's designed to suit the tool. This made it possible to adjust the tool normally for depth of cut. Using a plunge router is an asset because it can be preset for depth of cut, which also makes it possible to easily produce any number of cuts of similar depth.

Because of the mounting plate, there is some loss in maximum depth of cut; but when the original capacity is 2 inches, or more, losing ½ inch isn't crucial—even when forming mortises. If necessary, you can increase depth of cut capacity by mounting the router without its subbase and by substituting a thinner plate—like an ⅛-inch-thick aluminum plate.

Construction

While the project isn't difficult to build, it does have some crucial details that require special attention. Begin by studying the basic assembly drawing in FIG. 10-28 and by checking the material requirements listed in Table 10-1.

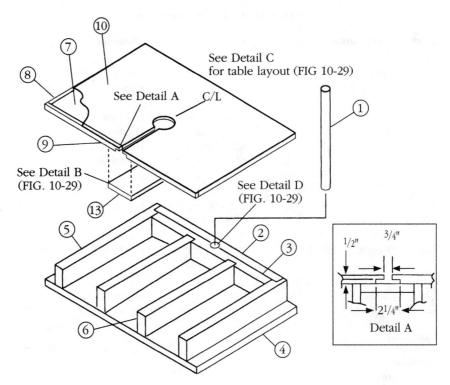

10-28 Construction details for the overarm's basic table.

You might have some trouble finding the thick-walled steel tube recommended for the column, but if you check the Yellow pages under "metal," you'll find a source, as I did—preferably one that deals with salvaged materials. Have the column and the router you plan to use on hand before you start cutting components to size.

Table 10-1 Material list for overarm router.

Key#	Part name	#Pieces	Size	Material
1	Column	1	1½" O.D. × 24"	Steel tube
2	Back-outside	1	1½" × 3½" × 24½"	Straight grain fir
3	Back-inside	1	1½" × 3½" × 23"	Straight grain fir
4	Base	1	¾" × 22" × 30"	Plywood
5	Rails	2	1½" × 3½" × 22"	Straight grain fir
6	Partition	2	1½" × 3½" × 19¾"	Straight grain fir
7	Table	1	¾" × 20¾" × 31"	Cabinet grade plywood
8	Trim	2	½" × ¾" × 20¾"	Pine
9	Trim	1	½" × ¾" × 32"	Pine
10	Cover	1	21¼" × 32"	Sheet aluminum
11	Arm	1	2" × 4" × 15¼"	Hardwood

Table 10-1 Continued.

Key#	Part name	#Pieces	Size	Material
12	Router support	1	½" × 5¾" × 19½"	Cabinet grade plywood
13	Slide support	1	¾" × 5" × 20½"	Hardwood
14	Inserts	*	¾" × 3" diameter	Hardwood
15	Pivot slide	1	¾" × 2¼" × 12½"	Hardwood
16	Filler	1	¾" × 2¼" × 5¾"	Hardwood
Fence				
17	Facing	1	1¼" × 3" × 21"	Hardwood
18	Guides	2	¾" × 3" × 12"	Hardwood
19	Stiffener	1	¾" × 1½" × 12½"	Hardwood
20	Height gauge	1	1" × 2½" × 2½"	Hardwood
Drawer				
21	Sides	4	¾" × 2⅞" × 18"	Plywood
22	Bottom	2	⅜" × 7½" × 18"	Particleboard
23	Back	2	¾" × 2⅞" × 6"	Pine
24	Front	2	¾" × 3¾" × 8½"	Pine

Hardware
(5) ¼" T-nuts
(3) ⁸⁄₃₂ T-nuts
(2) ⅜" Threaded inserts
(2) ⅜" × 1" Bolts
(1) ⁵⁄₁₆" × 2½" Carriage bolt w/washer & wing nut
(1) ¼" × 1" Wing bolt
(4) ¼" × 2¼" Threaded rod w/washers & wing nuts
(2) Drawer pulls (your choice)

Miscellaneous
T-nuts for inserts
⁸⁄₃₂ Screws for pivot points
#10 × 1½" FH screws
#10 × 1¼" FH screws
* As needed

Begin by cutting the two back pieces to length and then forming the dadoes in the front one for the partitions. Connect the two pieces with glue and No. 12 × 2½-inch flathead wood screws. It's important for the bottom edge of the assembly to be flat and square to adjacent surfaces. If it isn't, the 1½-inch hole needed for the column will not be true. Use a good spade bit or a Forstner bit in a drill press to form the hole, then test-fit the tube and check with a square on the front and side of the tube to be sure it is perfectly vertical. Drill for and install the ⅜-inch threaded insert for the bolt that secures the tube (Detail D in FIG. 10-29).

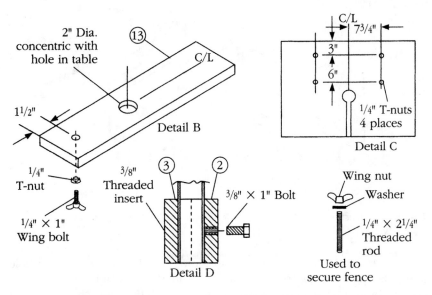

10-29 More construction details for the table. The four T-nuts in the table are needed for the threaded rods that secure the fence.

Cut the base, rails, and partitions to size and assemble all the parts with glue and No. 10 × 1½-inch screws driven up through the base.

Use a cabinet-grade birch or maple plywood when making the table. Cut it to size and then add the trim strips with glue and 4d finishing nails. Then attach the aluminum cover with contact cement. The aluminum cover can be considered an optional feature, or you can substitute a plastic laminate.

The next steps, forming and assembling the arm and the router support plate, deserve extra care—especially boring the hole through the arm. When boring the hole, the arm must be vertical and provide a snug fit for the column. The dimensions listed for the plate are suitable for the router I used. Before going further, check its length and front-end shape against the tool you will install. Bore the 2-inch hole so it will be on the plate's center-line, concentric with the router's chuck. After forming the rear hole for the column, attach the plate to the arm with glue and three No. 10 × 1½-inch flathead screws. Finally, form the kerf at the rear of the assembly and drill a through hole for the 5⁄16-inch carriage bolt (FIG. 10-30).

Secure the column in the base and slide on the arm assembly. Place the table, which butts against the column, so it is square to the substructure and its centerline mates with the center of the column. Mark the line lightly with a scriber if you have used an aluminum cover and then, with a ¼-inch plunge-point bit secured in the router, use the router to bore through the table exactly on the marked line. Remove the table and use a fly cutter or hole saw with a ¼-inch pilot to enlarge the hole to a 3-inch diameter.

The next sequence of steps requires the use of a table saw. First, cut the ¾-inch-wide cut that ends at the hole. With the blade projecting ⅜ inch, make repeat passes to form the T-shape for the pivot slide (see Detail A in

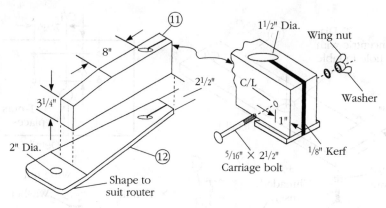

10-30 Construction details for the arm. A conventional nut can be substituted for the wing nut.

FIG. 10-28). Cut the slide support to size and, after forming the 2-inch hole and installing the T-nut (see Detail B in FIG. 10-29), attach it to the underside of the table with glue and several No. 10 × 1¼-inch flathead screws. The last table detail is to install the four ¼-inch T-nuts for the threaded rods that are used to secure the fence (see Detail C in FIG. 10-29). The T-nuts do not have to be set flush because the height of the drawers allows for clearance.

Coat all the top edges of the substructure with glue, place the table accurately, and then use clamps and whatever weights are handy around the perimeter to apply pressure at the midpoints. You can, of course, use screws to secure the table, but be certain they are set flush with the table's surface.

Accessories

Use a piece of wood for the pivot slide that is long enough so that after it is shaped, you can saw off a piece for use as the filler (FIG. 10-31). Three T-nuts, installed flush, are used in the slide for more flexibility when establishing the distance from pivot points to cutter. Pivot points, some long enough to penetrate workpieces, others short and pointed for use when a center hole isn't wanted, are made from ⁸⁄₃₂ screws.

It's best to use a fly cutter with a ¼-inch pilot drill to form the inserts because this will allow them to be sized for a snug fit in the table hole. Make several inserts, each for a particular purpose. Plug the hole in one insert with a dowel, install an ⁸⁄₃₂ T-nut in another for use with the pivot points, install ¼-inch, ⅜-inch, and ½-inch T-nuts in others for use with pins that can be made from bolts (see FIG. 10-31). The height gauge (also shown in FIG. 10-31), which is placed on the column between the arm and table, is optional but can be useful when you wish to return the arm to a specific height after making a special adjustment.

The fence

When making the facing for the overarm pin router fence (FIG. 10-32), form the wide dadoes for the guides before reducing the center area. The center

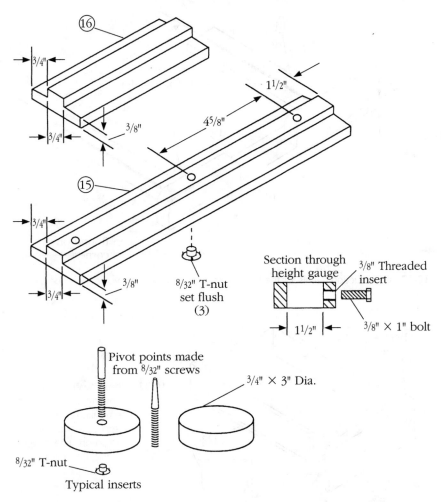

10-31 Construction details for the slide, inserts, and the optional height gauge. The filler (#16) is used to fill the gap in the table when the slide is not in use.

area must be wide enough so the router plate can sit on the ledge since there will be operations that make it necessary for the router bit to be as close to the table as possible. Attach the slotted guides to the facing with glue and No. 10 × 1½-inch flathead screws. The stiffener, which spans between the guides and butts against the back surface of the facing, can be installed with just glue.

Drawers

The drawers for the overarm pin router are assembled as shown in FIG. 10-33. This is a fairly basic design; nothing fancy, just practical and durable if assembled with care. Anyone with higher ideals can do more—for example, attach the front to the sides with dovetails instead of a prosaic rabbet joint.

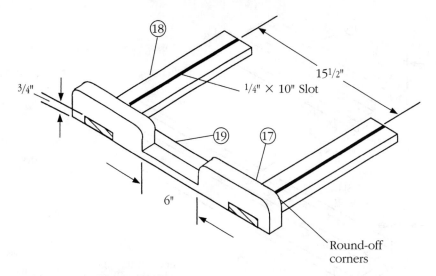

10-32 Construction details for the fence. Work carefully so the two arms will be parallel.

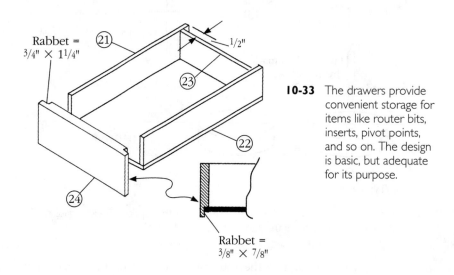

10-33 The drawers provide convenient storage for items like router bits, inserts, pivot points, and so on. The design is basic, but adequate for its purpose.

At work

Always be sure that the bolt that secures the column at its base and the carriage bolt that secures the arm to the column are tightened before beginning a job. If you find that the wing nut for the carriage bolt doesn't allow for sufficient tightening, substitute a nut that you can turn with a wrench.

Using a router with its base affords the same protection you would have when using the tool in normal fashion. If the router comes with a plastic shield, use it. As always, using hands close to a cutting area is ill advised. Be safety conscious—take precautions such as unplugging the tool before making any adjustments or changing cutters.

Typical applications

Straightforward routing is similar to shaping (FIGS. 10-34 and 10-35). The position of the fence determines the width of the cut; the height of the cutter above the table establishes its depth. Hold the work firmly down on the table and snug against the fence throughout the pass. Always move the work against the cutter's direction of rotation. Using an overarm router doesn't negate the good practice of making repeat passes to achieve deep cuts. The power of the router is, of course, a factor to consider, but, for example, if you need a ½-inch dado or rabbet and find that the tool labors, make one pass that is ¼-inch deep and then lower the cutter an additional ¼ inch for a second pass.

10-34 The fence is used for straight edge-cuts. The fence determines the width of the cut; the height of the bit above the table determines its depth. Repeat passes can be used for very deep cuts.

Mortising, whether the cut is contained within the boundaries of the work or open at an end, is done with the stock positioned as shown in FIG. 10-36. This is just another application where you appreciate the feature of a plunge router. The depth of cut can be established before the work is positioned, and the entry cut can be made simply by pushing down on the tool. Regardless of the type of mortise, it's a good idea to use a stop block to control the length of the cut; an especially useful setup when cuts must

10-35 When using the fence for edge shaping, the router bit might cut into the fence a little. The router mounting plate can serve as a hold-down, but its bearing must be light so the work will move smoothly.

10-36 The plunge router adds convenience and accuracy when doing mortising. Use a stop block (arrow) to gauge the length of cuts regardless of whether the mortise is contained or open.

be duplicated. Remember that mortises formed with a router bit will have semicircular ends; therefore, the tenon must be rounded off to suit.

Dovetail slots are formed with the tool and the work positioned as they would be for mortising (FIG. 10-37). If you find it difficult to produce the shape in one pass, make an initial cut with a straight router bit as if you were forming a simple groove. Then install the dovetail bit and make a second pass. If you work this way, the dovetail bit will have little material to remove.

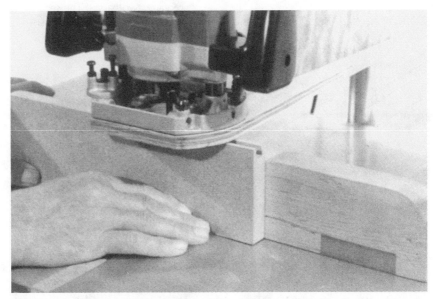

10-37 To form a dovetail slot in the edge of the workpiece, hold the work firmly and move it slowly.

To create the *dovetail pin*, two passes are required (FIG. 10-38). Once the setup is established, producing the second part of the shape is a matter of turning the stock end-for-end and making a second pass.

Freehand routing is accomplished by removing the fence and working with router bits that have an integral or ball-bearing pilot (FIG. 10-39). Ball-bearing pilots are recommended because they rotate independently of the cutter and thus eliminate the friction—and possible burn marks—that is characteristic of an integral pilot. In either case, it is crucial, for accuracy and safety, for the work to have sufficient bearing surface against the pilot. When the thickness of the work itself isn't enough for a good bearing edge, attach a piece temporarily to the bottom of the work to compensate.

Pivot routing uses special inserts and pivot points (FIG. 10-40). The point you choose depends on whether or not the work can have a center hole. Mount the work on the slide, then lock the slide after establishing the correct distance from pivot point to cutter. Begin your cutting by rotating the work in a clockwise direction (FIG. 10-41). Use an insert that has a central hole when you need to cut through the stock.

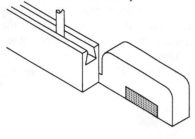

Dovetail "slot"
made in one pass

10-38 The matching pin for the dovetail requires two passes. The second one is made after the stock has been turned end-for-end.

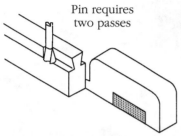

Pin requires
two passes

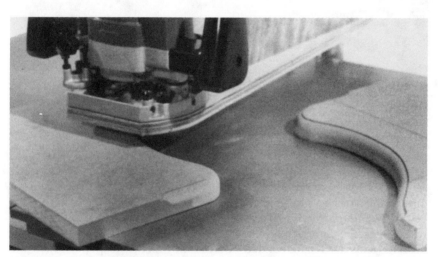

10-39 Shaping curved edges is done freehand with the work guided by the pilot on the router bit.

Pattern routing requires an insert with a post that is concentrically aligned with the router bit. A pattern, sized to compensate for the guide post, is tack-nailed to the bottom of the work (FIG. 10-42). The operation involves maneuvering the work so the pattern bears constantly against the post (FIG. 10-43). The relationship between pattern and post is shown in FIG. 10-44. The pattern has sufficient bearing against the post; the projection of the post above the table is less than the thickness of the pattern.

10-40 *Points*, used for pivot routing, are situated in the slide. *Posts*, used for pattern routing, thread into T-nuts that are installed in the inserts.

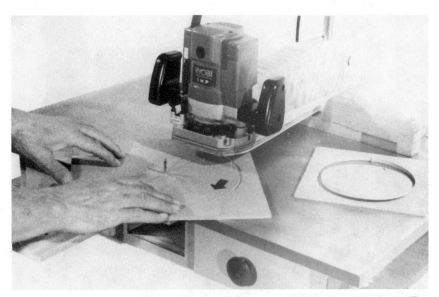

10-41 When pivot routing, the work is impaled on a point and rotated clockwise. The position of the slide determines the radius of the circle.

Straight line *surface carving* is done by using the fence to guide the work (FIG. 10-45). In FIG. 10-45, a straight, round end router bit is being used to form semicircular grooves, but there are many options. Pilotless-type bits must be used for this application.

10-42 An insert, fitted with a post (arrow), serves as a guide when doing pattern routing. The pattern is attached to the underside of the workpiece. It's a good idea to form the hole for the post by using a plunge bit in the router. This will assure concentricity.

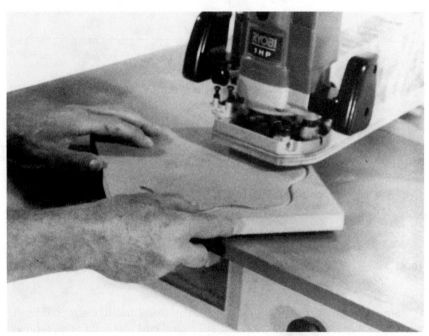

10-43 Moving the work so the pattern bears constantly against the guide-post allows the cutter to duplicate the pattern's shape.

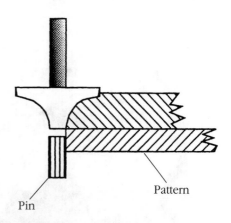

10-44 When pattern routing, the height of the guide post above the table must be less than the thickness of the pattern. Work is guided by the pattern that bears against the guide pin.

Pattern

Pin

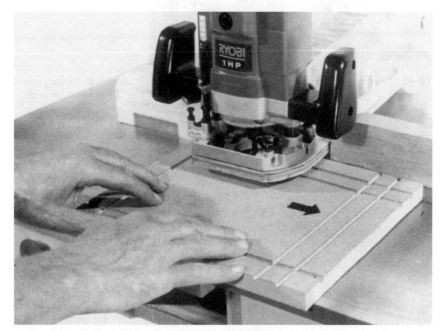

10-45 When making surface cuts that intersect and when rabbeting or shaping adjacent edges, make the cross-grain cuts first. Final cuts with the grain will remove imperfections that are likely to occur at the end of cross-grain cuts.

DOWEL TENONING JIG

Reducing the end of a dowel, or *round*, is a practical way to form integral tenons on rungs, stretchers, some leg designs, and similar components. There are various methods that can be used, for example, turning the dowel in a lathe or using plug cutters. The method I use most often though is to complete the task using the router-mounted jig shown in FIG. 10-46. The idea is straightforward; the dowel is held between V-blocks and turned slowly as it is advanced over the cutter. The adjustable stop controls the length of the tenon.

10-46 This dowel tenoning jig is attached to the base of a router. A secondary use is to form particular size dowels. The tenon is cut off after it has been formed.

It's important, of course, to secure the router. A flat top on the tool helps but in any event, a large *handscrew,* which in turn is clamped to the workbench, will hold the router still. The screws that hold the top V-block are tightened enough to steady the round while still permitting it to turn. Turn the work slowly as you move it forward. Get to the final diameter by making repeat passes when it's necessary to remove a lot of material. You'll find, if you don't rush, that a final swipe with sandpaper will produce a perfectly true tenon.

Construction details for the dowel tenoning jig are offered in FIG. 10-47. Drill the holes for the steel rods so they will be a tight fit in the stop. The holes for the rods in the main part of the jig should also be a little under-size so the stop will maintain its position without the need of additional set screws. Incidentally, there is no reason why the jig can't be attached to a router/shaper table. If so, skip making the stop; the table's fence will serve in its stead.

PANEL ROUTING JIG

The panel routing jig, shown being used in FIG. 10-48, is a practical accessory that can be used for a variety of template-guided routing operations. For example, the panel routing jig can be used to form inlay lines in small projects as well as to create decorative designs in cabinet doors and furniture components like the front part of a drawer. A feature of the project is its flexibility. It is easily organized for working on parts that are square or rectangular in any size up to the maximum opening of the jig, which is about 20½ inches × 20½ inches.

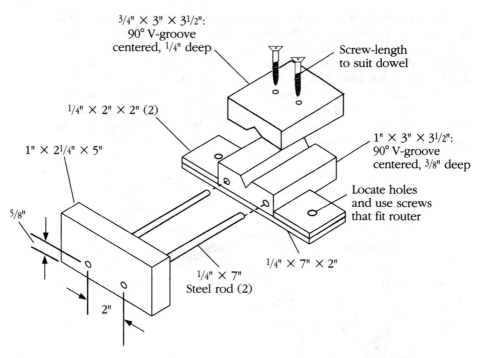

3/4" × 3" × 3 1/2":
90° V-groove
centered, 1/4" deep

Screw-length
to suit dowel

1/4" × 2" × 2" (2)

1" × 3" × 3 1/2":
90° V-groove
centered, 3/8" deep

1" × 2 1/4" × 5"

Locate holes
and use screws
that fit router

5/8"

1/4" × 7" × 2"

1/4" × 7"
Steel rod (2)

2"

10-47 Construction details for the dowel tenoning jig. The main part of the jig can be used on a router/shaper table.

10-48 The panel routing jig is used with custom-designed templates to incise decorative grooves in, among other things, cabinet doors and drawer fronts.

The jig consists of four frame pieces that are put together as shown in FIG. 10-49. The hardwood part that extends from one end of each component slides in the "slot" that results when the pieces are assembled. Work carefully when cutting and assembling this jig because it will work as it should only if ends and edges are perfectly square. When the hardwood extension on one part is inserted into the slot of an adjacent frame, remember that the angle between them must be 90 degrees. Figure 10-50 offers an alternate, possibly more durable way to design the tightening system. The bases of the modified T-nuts must not be more than ¼ inch.

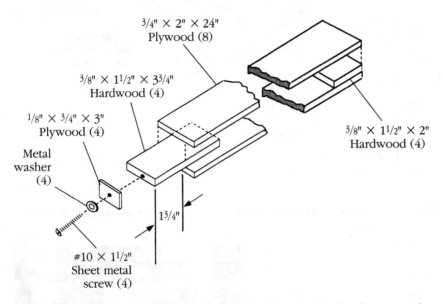

3/4" × 2" × 24"
Plywood (8)

3/8" × 1¹/2" × 3³/4"
Hardwood (4)

1/8" × 3/4" × 3"
Plywood (4)

Metal
washer
(4)

3/8" × 1¹/2" × 2"
Hardwood (4)

1³/4"

#10 × 1¹/2"
Sheet metal
screw (4)

10-49 Construction details for the panel routing jig. Note that the four frame parts for the jig are identical. Ends and edges must be square if components are to mesh correctly.

The jig is used with templates that you design. They can be any shape you choose but each of them must have a ⅜-inch × ⅜-inch tongue on the edges that will be inserted in the slots of the frame parts (FIG. 10-51). As shown in FIG. 10-52, one set of templates can be organized to produce various designs.

After the templates are in place, and the jig is secured to the work with clamps or, maybe, double-face tape, then the router, with template guide and bit in place, follows the guide lines that are established by the jig and the templates (FIG. 10-53). You may find on large projects that the weight of the router causes some deflection in frame parts. Counter this by using pieces of ⅜-inch stock in the slots, as shims, wherever they are needed. If you have made enough templates, the idle ones can be used instead of extra shims.

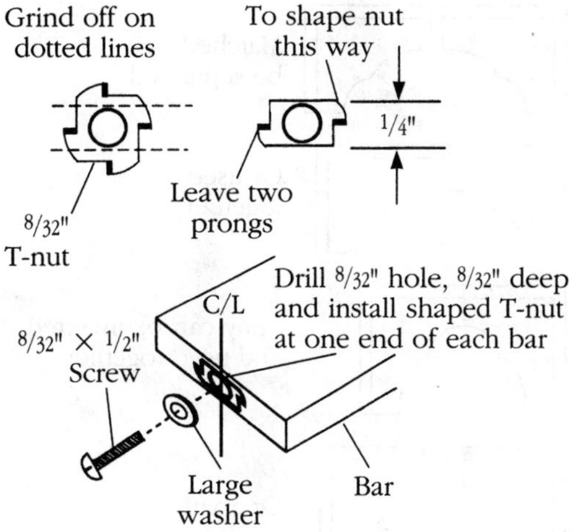

Grind off on
dotted lines

To shape nut
this way

1/4"

8/32"
T-nut

Leave two
prongs

10-50 An alternate method
for securing the panel
routing jig frame
components.

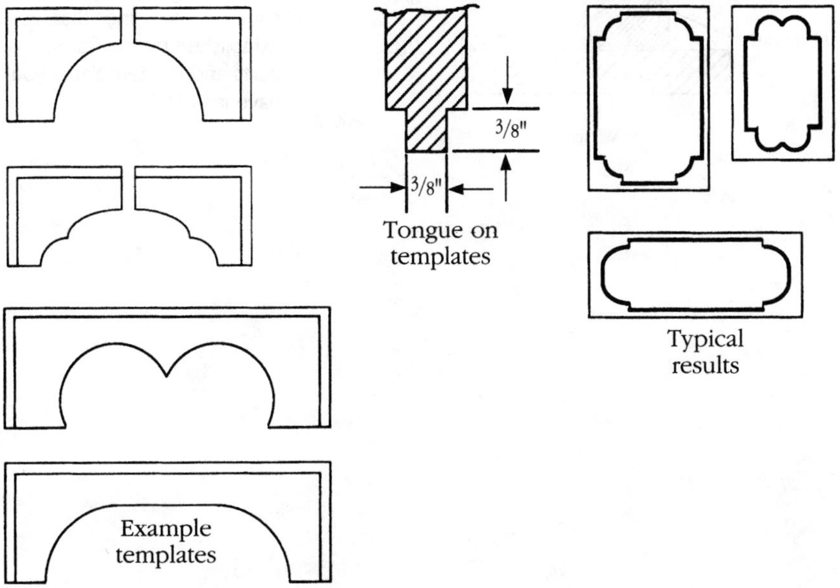

C/L

Drill 8/32" hole, 8/32" deep
and install shaped T-nut
at one end of each bar

8/32" × 1/2"
Screw

Large
washer

Bar

3/8"

3/8"

Tongue on
templates

Typical
results

Example
templates

10-51 Templates for the panel routing jig can be any shape you choose, but they must
have tongues that fit the slots in the frame members. Pictured are some examples of template designs.

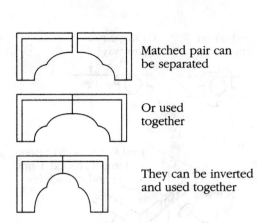

Matched pair can
be separated

Or used
together

They can be inverted
and used together

Or apart

10-52 How one set of
templates can be
organized to produce a
variety of designs.

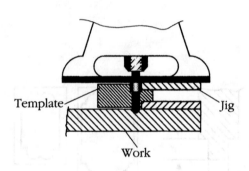

Template

Jig

Work

10-53 Panel routing does not differ
from standard template
routing procedures.
Template guide in the
router follows lines that are
established by the frame
parts and the templates you
have installed.

11

More router applications

The more familiar you become with the router, the more useful and interesting applications you will find for it. The procedures that follow are part of the router story.

PATTERN ROUTING

Although the words *template* and *pattern* are often used interchangeably, there is a basic difference between the two. Essentially, a template is a guide that is made to be followed by the router when it is equipped with an accessory template guide. Conversely, a pattern might be an actual project component or a specially made unit with which the router, generally with a piloted bit, can be used to form one or more duplicate pieces. The pattern could also be shaped to provide just part of the configuration that is required on the workpieces (FIG. 11-1).

When the pattern is a complete unit used to produce duplicate parts, it is the same size and shape as the pieces needed. When the pattern is used for a partial cut, say a particular curve on one edge of a component, then it can be oversized in unimportant areas so that, if for nothing else, it can make the pattern easier with which to work.

Workpieces are oversize to begin with and can be attached to the pattern in various ways—by clamping, spot gluing, or tack-nailing (FIG. 11-2). Figures 11-3 and 11-4 show the type of bits that can be used for pattern routing. Because piloted bits are used, the edges of the pattern must be smooth and flawless.

11-1 The pattern indicated by the arrow can be the full size and shape of the project, or it can be made to supply the shape for a limited area.

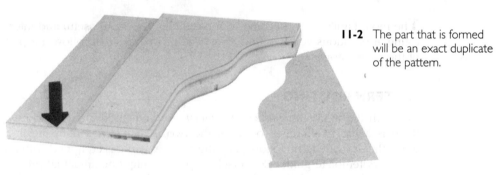

11-2 The part that is formed will be an exact duplicate of the pattern.

11-3 Straight bits with integral or add-on pilots can be used for pattern routing.

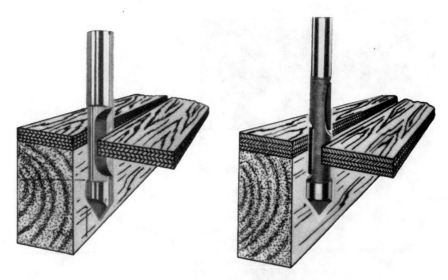

11-4 These pilot panel bits can penetrate like a drill and are also used for pattern routing. They are especially useful for making internal cutouts. Both can be used for plunge cutting. The one on the right is a *stagger tooth* design.

PIERCING

Piercing is a term usually associated with the scroll saw and straight cutting tools, like the table or radial arm saw, but there are several ways the router can be used to do similar operations. Piercing is often used to create small or large decorative panels, like those displayed in FIGS. 11-5 and 11-6.

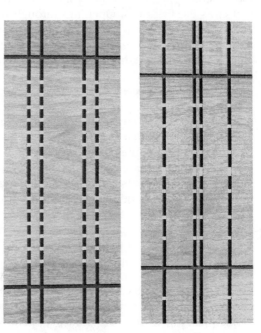

11-5 This type of piercing calls for making cuts on both sides of the stock. Openings occur wherever the cuts cross because depth of cut is a bit more than half the stock's thickness.

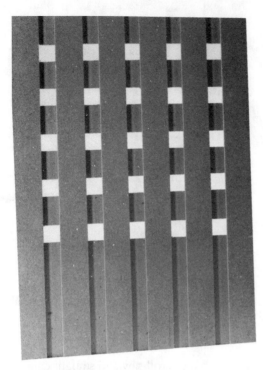

11-6 Various effects can be created by the number, position, spacing, and width of the cuts. Using a smaller bit on one side will result in rectangular openings.

The idea is simple. Set the projection of the router bit to a little more than half the thickness of the stock. Make a series of cuts on one surface of the material. Then, invert the workpiece and make a second series of cuts that cross the first ones. Openings through the work appear wherever the cuts cross. Variations, which are limitless, are affected by the following factors: the number and spacing of the cuts; whether the cuts cross at right angles or obliquely; and the shape of the bit. Some very intriguing patterns are created when pointed decorative bits are used, but be sure the thickness of the stock permits necessary depth of cut. Another idea is to work with a pivot guide to make circular grooves (with various center points) or to make straight or angled cuts on one surface and circular ones on the other side. It's also possible to make stopped or blind cuts for trivet-type projects like the store-bought item shown in FIG. 11-7.

The shape of the openings depends on how you plan the cutting. Similar, equally spaced cuts that cross at right angles create square openings. If the cuts cross at oblique angles, then the openings appear as diamonds. Arc-type openings result if you make wide, straight cuts on one side and circular grooves on the opposite surface. Experimenting to check results can waste material; it's best to preview possibilities by working on paper with a straightedge and a compass.

Another type of piercing work, created with through cuts rather than grooves, is demonstrated in FIG. 11-8. The example shows how a pivot guide can be used to form perfect circular openings or discs, but the system also works when a pattern is used. Straight, pilotless bits are used when the router

11-7 Stopped or blind cuts can also be used for pierced projects. Here the technique was used for a trivet.

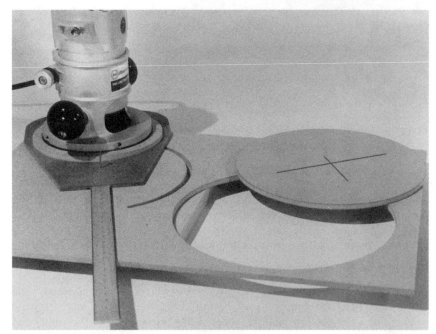

11-8 Piercing also applies to operations like this. Because the bit passes through, the work must be elevated on outboard support blocks or placed on a scrap piece of material.

is directed by something like the pivot guide. When piercing by following a pattern, use plunge-type bits like those that were shown in FIG. 11-4. Figure 11-9 shows how piercing can often result in two components, although not necessarily for the same project.

HOLLOWING

Hollowing applies to the contained recesses required for the store-purchased products shown in FIG. 11-10. The concept is simple: to reduce the

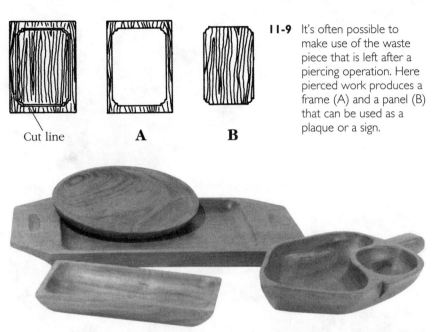

Cut line   **A**   **B**

11-9 It's often possible to make use of the waste piece that is left after a piercing operation. Here pierced work produces a frame (A) and a panel (B) that can be used as a plaque or a sign.

11-10 Projects like these store-bought servers are formed by using hollowing procedures. In many cases, it's best to shape perimeters after the cavities are formed.

thickness of stock between perimeter areas so cavities of optional depth and shape will be formed. Because the router has to be moved in a rather erratic cutting pattern to cover all areas, the tool should be equipped with a special subbase to adequately span across the outside edges of the project and provide good support for the tool regardless of where it is cutting (FIG. 11-11).

How deep to cut on a single pass depends, as always, on the horsepower of the tool, but a light-duty model can accomplish as much as a heavy-duty machine simply by repeating cut patterns after increasing bit projection. To create informal or rustic results, the outline of the cavity can be followed by moving the router freehand along marked guidelines. Any slight deviations from the marked lines only contribute to the appearance you envision.

If you want a more formal appearance—straight lines rather than wavey ones—use a template/template guide setup to outline the cavity. The template does not have to be elaborate. Strips of wood with straight edges or curves, depending on how the cavity must be shaped, will serve as a template when they are tack-nailed to the project material (FIG. 11-12).

There is another factor to consider. If the router is equipped with an extra-long subbase and is to be used with a template guide, then the auxiliary subbase must accept the guide so that only the sleeve of the guide projects (FIG. 11-13). This calls for a normal hole for the guide to pass through, plus a counterbore so the accessory is seated flush with the bottom surface of the base when it is installed.

11-11 For most hollowing jobs, it's necessary to equip the router with a special subbase so the tool can be supported over the entire cut area.

11-12 It's often possible to create a template for outline cuts simply by tack-nailing strips of wood. The strips can have curved edges as well as straight ones.

There are various ways to accomplish this, the easiest depending on the workshop equipment you have. If you have a set of Forstner bits, you can use one size to drill a hole of limited depth (actually a circular groove) for the counterbore and a second size for the through hole. The same procedure can be followed if you have a set of hole saws. In each case, counterbore first. These are quickie methods. If you lack the equipment, you can achieve the same results by using the router and straight bits. Make one template with a hole to suit the outside diameter of the counterbore and a second one for the through hole.

Cutting is started by following the template to outline the cavity and then moving the router between outline cuts to remove waste material (FIG. 11-14). You can make parallel cuts, moving the router in ideal fashion so the bit is cutting into the work, or you can move the router erratically. Chances

11-13 A counter bore as well as a pass-through hole must be formed in the special subbase in order for the template guide to seat correctly.

11-14 Start a hollowing job by making outline cuts and then moving the router about to clean out waste.

are that you will use both cutting procedures. Just be sure to keep the sub-base flat on the surfaces of the template and that you make repeat passes, when necessary, to achieve the full depth of cut you need. This procedure works fine for partial cavities as well as total ones (FIG. 11-15).

11-15 Partial cavities result when you tack-nail a template to an inside area.

A special type of hollowing procedure is required when the cavity must slope in a particular direction. This applies to project components like chair seats (FIG. 11-16) or to units like cutting boards where you might want to channel meat juices to a collection area. The answer is to make a profile template like the example in FIG. 11-17. Because the forward edge of the template is higher than rear areas, the router will cut less toward the front and progressively deeper as it approaches the template's back bearing edge.

Make the first pass, following the template's edge, with a large round-bottom bit. This develops the outline of the shape you need and forms a neat cove around the base of its perimeter. Then switch to a large-diameter straight bit to remove the waste between the cove cuts. A special subbase that is long enough to support the router in whatever area it is working on will be needed.

Whether you can shape top edges of cavities with the router will, to some extent, depend on the depth of the cut. There is no problem on deep cuts because a piloted bit can be guided by the shoulders of the cavity (FIG. 11-18). Solutions must be based on the particular facets of the project at hand. It might be possible to work with an edge guide that rides the outside

11-16 Chair seats are often tapered from front to back. This is a hollowing job that is accomplished with a special template.

edges or by making a special secondary template that can be attached over the area or areas that are already formed.

There is no all-purpose solution. This is where router wisdom enters the picture. You examine the situation, review other router procedures and then attempt to arrive at the best method for getting the job done. You'll be surprised, as you go along, how you will start thinking "router fashion."

LEVELING AND SURFACING

After boards have been glued together as slabs for tabletops, workbenches, cabinet sides, and similar project components, it is often necessary to work with tools, like scrapers, hand planes, or sanding machines, to bring the assembly to a true level surface. The extra steps are needed, not because of carelessness when gluing and clamping, but because individual parts usually have flaws like a crossgrain cup or a lengthwise bow. No matter how minor the defects, they can't be eliminated by clamping. The end result is a less-than-perfect surface.

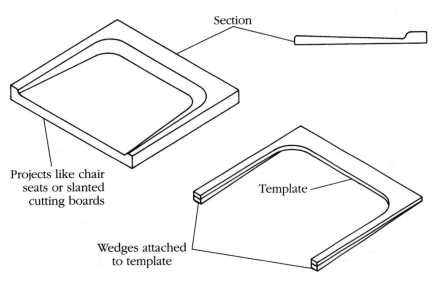

11-17 The template for a chair seat, or similar projects, should look like this. The router cuts progressively deeper as it approaches the back edge.

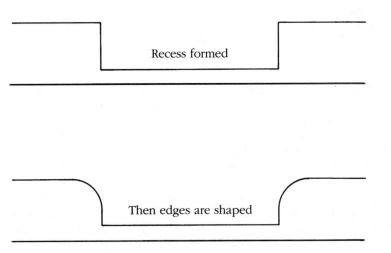

11-18 Inside edges of the chair seat can often be shaped with the router after hollowing is complete.

You can rectify discrepancies by working with a hand plane, with a belt sander, or by using both tools, but results will be better and easier to achieve if the job is done with a router. By working with the jig that is shown in FIG. 11-19, you'll have mechanical control of the operation so the possibility of high and low spots, that can easily occur when planing or sanding, will be eliminated.

The jig is a track assembly that spans across the work and rides on strips that are temporarily attached to the slab with a few nails. The top

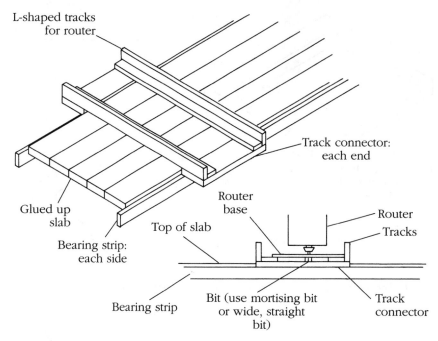

11-19 Construction details for a slab leveling jig. This jig makes it fairly easy to smooth and level surfaces of new or existing projects.

edges of the bearing strips (FIG. 11-19) should be below the surface of the slab by a bit more than the thickness of the track connectors. Setting the strips correctly is an important part of the operation. Check with a level to be sure their top edges are on the same plane so that when the jig is placed, it will be at the same height over all areas of the work.

Work with the largest straight bit or mortising bit you have and adjust its projection so it will just touch the work at its lowest point. Cutting is done by moving the router to and fro over the slab as the jig is moved from one end of the slab to the other. The distance between the tracks should be an inch or so greater than the diameter of the router's base. This allows you to make a few overlapping passes each time the jig is shifted.

A jig of this type doesn't have to be limited to new construction. It is just as practical for renovating an existing surface, like the top of a much-used workbench. After it has served its primary purpose, the jig can still be used for router jobs like cutting straight grooves.

SURFACING SLABS

A jig that is similar to the one mentioned under "Leveling and surfacing" can be used to flatten the ends of small logs that you wish to mount in a lathe or use as a pedestal. This jig can also create a smooth surface on a rough slab that will be used as the top of a bench or coffee table. In both cases

(FIGS. 11-20 and 11-21), a special cage or holding jig must be made to suit the size and shape of the material.

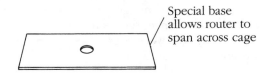

Special base allows router to span across cage

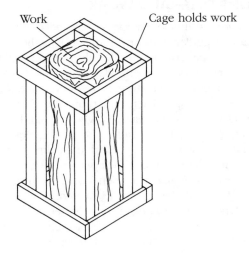

Work

Cage holds work

11-20 Flattening the ends of logs called for a special *cage* that holds the work securely in vertical position.

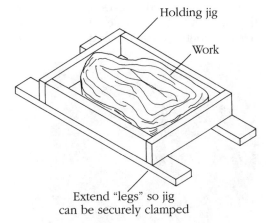

Holding jig

Work

11-21 A similar frame setup is used to secure slabs for cutting.

Extend "legs" so jig can be securely clamped

A cage can be made oversize if you contemplate doing a lot of this type work. With an oversize cage, wedges or pieces of scrap can be used to keep the work secure. The same idea holds true for a holding jig because various size workpieces can be secured by impaling them on nails driven up through the base of the jig, or by using scrap pieces between the work and the jig's sides.

The router must be equipped with a special subbase that is more than long enough to span across support areas of the jigs. The auxiliary bases do not have to be fancy; long, straight boards with a hole for the bit to pass through are good enough. The thickness of the board dictates whether you must use attachment screws that are longer than the standard ones. Results are best and the work goes fastest when you cut with the largest straight bit or mortising bit available.

SOME FREEHAND WORK

Freehand router operation simply means that cutting is done without mechanical guidance of any sort—no straightedges, no templates or template guides or patterns, no jigs. Results depend totally on how the operator "sees" and how he or she manipulates the tool. It's an area of router work that is very intriguing, but one whose importance relates directly to user interest and work scope. There are professional sign makers, wood-carvers, and wood sculptors, for example, who can handle a router with enviable artistry. They can carve letters and numbers of exclusive design and create bas-relief plaques or figures in the round in an almost matter-of-fact way. But any one of them will tell you that the apparent ease with which they accomplish such work is misleading. The talent comes with practice—a lot of it.

One of the problems to overcome when doing freehand work is the tendency of the router bit to volunteer a cut direction and to follow grain lines. When practicing, it is wise to select a material that presents the fewest obstacles to smooth cutting. For example, soft pine or redwood are both much easier to work with as opposed to wood species that have hard and soft areas combined, like Douglas fir. A test project can be something like the initial in FIG. 11-22, where the appearance is rustic and a few irregularities contribute to rather than detract from the project's appearance.

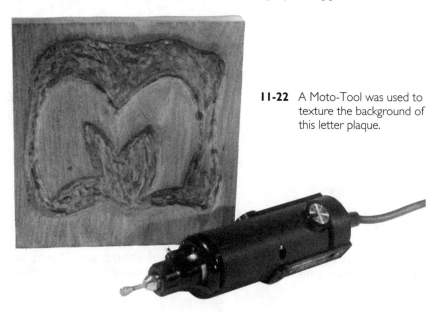

11-22 A Moto-Tool was used to texture the background of this letter plaque.

Another test possibility is recessing the background of a figure that has already been accurately outlined using a template (FIG. 11-23). You can still get careless here, but the possibility of cutting into the figure is minimized because of the outline groove. If you work with a straight bit you can achieve a smooth, flat background. If an interesting texture is what you prefer, you can use a round end or veining bit. On freehand routing projects, it's best to start cutting at the perimeter of the work so that as you move toward interior areas, the router will be supported by the figure. In all cases it's important for the tool to be held on a level plane.

11-23 Recessing the background of a figure that has been outlined with a template is a way to practice freehand routing.

When working freehand, it's often necessary to break the standard feed direction rule; that is, moving the router so the bit cuts *into* the work. The grip on the tool should be firmer than ever, and feed speed should be lessened. I find that it's sometimes practical, especially when following a line, to deliberately break the rule. The bit will tend to move away from the line rather than cross it.

Practice sessions should include quite a bit of cutting obliquely to or across to the grain. When you can make cuts like this while accurately following a line, you'll be well on your way to freehand expertise. Start with small bits and shallow cuts. Equip the router with a subbase that has an oversize hole for the bit or, better still, install a see-through subbase so you'll be better able to see the layout lines. Another secret of good freehand work is anticipating how the router should be moved to follow the lines that are ahead. This is easier to do with the see-through subbase.

CARVING WITH A ROUTER

Freehand work doesn't have to be limited to bas-relief projects. If you view the router without its base, you can see that it can be gripped somewhat like the handpiece on a flexible shaft or used like a high-speed grinder. This particular application is more feasible with small routers or, preferably, with the motor of a laminate trimmer. Hand-holding the motor of a big, 3-horsepower unit is not suggested; however, if the motor is in a fixed, stable position to apply the work to the cutter, not the reverse, this might be an alternative method to use.

One way to secure a motor is with the type of jig that is shown in FIG. 11-24. Because of its extended base, the jig can be clamped securely to a workbench so that you can apply the work to the cutter (FIG. 11-25). *Don't attempt to do this kind of work with standard router bits!* They can dig in

11-24 A jig of this type can be used to secure the motor of a router or laminate trimmer for jobs that can be done more conveniently by hand-holding the work.

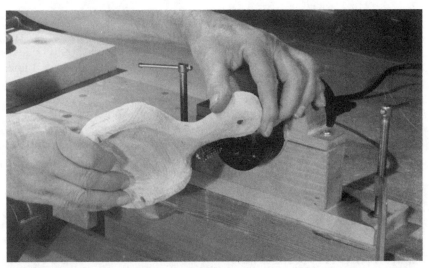

11-25 Shaping "in the round" is feasible with a router and a rotary rasp cutter when the work can be done this way.

and pull the work from your hands no matter how carefully you work. Furthermore, it's dangerous for your hands to come anywhere near those cutting edges. Even with the type of cutters that should be used for this application, like rotary burrs, mills, files, and such, great care should be taken with hand placement. These tools don't cut like router bits, but they do cut! Be sure the cutters that you use have the correct shaft diameter for the collet and that they are specified for safe use at the speed and horsepower of the router's motor.

Construction details for a typical motor-holding stand are shown in FIG. 11-26. To make one, follow the instructions that were outlined for the Peripheral Jig in chapter 10. The only difference between the two units is in the shape of the base.

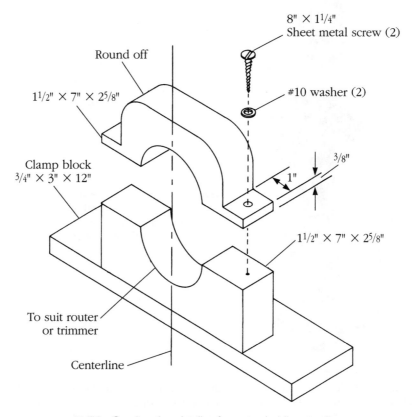

11-26 Construction details of a motor-holding stand.

Projects "in the round," like the sculpture shown in FIG. 11-27, can be accomplished partially by working with the router in a fixed position and partially by clamping the work and applying the router to it. It's often possible to minimize the amount of waste that must be removed with cutters by presawing the stock to its basic form (FIG. 11-28). If you have a band saw, this kind of preparation can be done by using the technique called *compound*

11-27 Wooden sculptures like this can be accomplished entirely with a router motor. The motor can be secured in a stand or, with the work secured, used like a high-speed grinder.

11-28 The amount of waste you must remove when carving figures and similar projects can be minimized by preshaping the workpiece.

sawing, which is often used to shape project components like the cabriole leg. The profile of the work is drawn on two adjacent sides of the work. After one profile is sawed, the waste pieces are tapped back in their original positions to give the workpiece a flat surface to ride on when the second profile is cut (FIG. 11-29).

LETTERS AND NUMBERS

Letters and numbers of just about any size and shape can be formed with a portable router, either by using the tool freehand or by working with templates. It takes practice to get optimum results when working freehand, but

11-29 Compound sawing on a band saw is a quick way to preshape stock for carving. When pieces are small, say in the 2-inch-square area, you can do the same kind of work with a jigsaw.

the method allows a good deal of liberty in terms of design. Freedom of expression is the prime reason for working without guides. Any style of numbers and letters you sketch on the workpiece, whether standard or original, can be carved out with the router. Professional signmakers often work freehand because they can turn out a one-time product that is especially right for a particular establishment.

Workers who might make a set of house numbers or a name plaque just once can do the chore more easily by working with templates. Commercial templates are available, but there are ways to work that reduce costs and that don't limit you to the particular letter and number style that is offered. The letter plaque shown in FIG. 11-30 was made by using the letter "E" from a store-bought cardboard sign as the design for a template. Hardware stores and home supply centers carry many cardboard or plastic ready-to-use signs and sets of numbers in various sizes and styles. It might be hard to find a sign that contains all the letters you need, but there are also complete alphabets with letters, as well as numbers, available individually (FIG. 11-31).

Another way to acquire a variety of different shaped letters and numbers for template material is to use headlines and titles in newspapers and magazines (FIG. 11-32). They can be enlarged at a copy shop to the size you want. There might be the problem of finding all the letters you need in a particular style, but you can fill in by drawing missing ones. If nothing else, printed material can be a source of type styles.

You can also be as creative as you please by designing your own alphabet and numbers. The examples shown in FIG. 11-33 were drawn with draft-

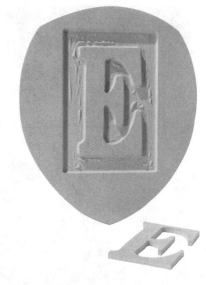

11-30 Letters or numbers in various sizes and styles can be used as patterns to create templates for router use. Notice here that the background cutting has left irregular ridges that contribute to the motif and make the cut seem deeper than it really is.

11-31 Ready-made letter and number designs can be purchased in sets or as individual pieces. They are made for use as is, but you can view them as patterns for templates that can be followed by a router.

ing instruments. Tools you can use include compasses, circle templates, French curves, and so on. When letters or numbers are to be made as individual units and are designed somewhat along the lines of those in FIG. 11-33, much of the waste material can be removed before routing by boring holes.

Whichever way you go, the pattern for the letter or number is attached to the template material with tape or an adhesive, like rubber cement, and the figure is sawed to shape (FIG. 11-34). The sawing can be done on a jigsaw, with a coping saw, or, if you choose to work freehand, with the router. At this point, because you already have the letter, you might ask—why go further? It depends on how you wish the project to appear and whether you wish to save the unit for possible future use. Formed letters can be overlaid on a shaped plaque or a plain board, or used as guides for bas-relief work.

11-32 Titles and headlines from magazines and newspapers are another source of letter and number designs.

11-33 You can create your own style of letters and numbers by working with drafting tools like circle templates and French curves. Dotted lines in figure indicate areas of letters and numbers that can be formed by boring holes to remove waste material.

Figure 11-35 shows one arrangement that allows for recessing the background to "raise" the letter, which is the technique used for the plaque in FIG. 11-30. The template and the straight pieces that are guides for the outline cuts are tacked-nailed or spot-glued to the workpiece. The first step is to move the template guide along all edges to form outlines for the figure and frame. Then work on removing the material between the first cuts. When making a sign, the frame pieces are moved for proper spacing after each figure is formed.

You can often save time and effort when making individually shaped letters or numbers by using the following technique. Do the routing on stock that is twice or even three times thicker than the thickness you need for the parts. Then resaw to get exact duplicates (FIG. 11-36). The idea is practical because a letter or number sign often needs more than one figure.

11-34 The letter or number that you have selected or created is attached to the template material that is then sawed to profile shape. This can be done on a jigsaw, a band saw, and by hand with coping saw.

11-35 After the template is formed, it is attached to the workpiece by tack-nailing or spot-gluing. The arrangement shown here, with strips of material used as a template, provides an outline cut.

Figures 11-37 and 11-38 show some ideas you can consider for signmaking. Incised figures (FIG. 11-37) can be filled with a contrasting material like a wood dough or a plastic metal. A script-type lettering (FIG. 11-38), accomplished by piercing with the router, can be overlaid on a contrasting backing.

Anyone who wants signmaking to be a fairly straightforward procedure or who anticipates a lot of work in this area either for fun (making signs as gifts) or for profit, should check out an accessory from Sears called the Rout-A-Signer (FIG. 11-39). This unit comes complete with alphabets, sets of numbers, and a carousel to store the stencils systematically and conveniently (FIGS. 11-40 and 11-41).

The accessory works something like a pantograph. The letter or number template is secured in a special holder at one end of the machine. A sty-

11-36 The piercing technique can be used to form letters or numbers. You can produce a number of parts by doing the initial cutting on stock that is thick enough to be resawed into separate, duplicate pieces.

11-37 Figures that are recessed will have more impact if the grooves are filled with a contrasting material.

11-38 Another idea is to use the router freehand to create script-type lettering and then adhere the project to a contrasting background.

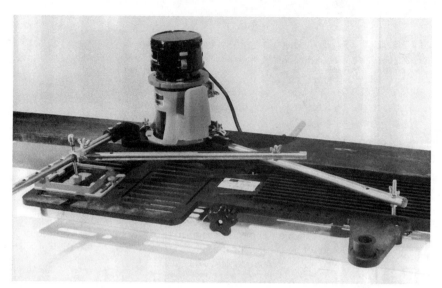

11-39 The Sears Rout-A-Signer should interest anyone thinking of producing letter or number signs in quantity.

11-40 How the tool is arranged allows you to produce letters or numbers ranging from 1 inch to 4½ inches with the one set of stencils (or templates) that are supplied with the unit.

lus is moved to follow the bearing edges of the template (FIG. 11-42). The router, mounted on a special subbase, duplicates the motions by means of an arrangement of tubes and bars that have specific swivel points. Essentially, it's how you select the swivel points that makes it possible to use the jig for letter and number sizes that range in height from about 1 inch to 4½ inches, while still using the templates that are supplied with the unit. The

11-41 The carousel of the Rout-A-Signer stores stencils conveniently. Symbols on its perimeter make it easy to select a particular letter or number.

11-42 A stylus is moved to follow the bearing edges of the templates that is secured in a special clamp-type fixture at one end of the jig. Because of a pantograph-type action of the accessory, the bit in the router follows the movements of the stylus.

workpiece, by means of an adjustable integral clamping arrangement, is secured at the rear of the accessory and moves longitudinally for spacing between figures. The correct spacing is assured if you follow the instructions supplied with the jig, which is the answer that makes optimum results with accessories a foregone conclusion.

Black & Decker offers another system for accurate routing of nameplates, signs, and house numbers (FIG. 11-43). The Black & Decker kit includes complete sets of 1¾-inch and 2½-inch size letters and numbers, and a holding device that has a built-in clamp for the templates. The workpiece is clamped securely; the template holder is moved at a spacing of your choice for each of the figures you cut. The concept calls for using a router that is equipped with a template guide. Working with V-groove bits or straight bits is a standard procedure. Any router that you own can be used with router guides of this nature. How deep the figures will be incised is optional. If you arrange the work in an elevated position and work with a straight bit, you can *pierce* the figures right through the workpiece.

11-43 Black & Decker's idea for accurately producing letters and numbers in sizes of 1¾ and 2½ inches consists of templates secured in a special holder that can be clamped anywhere on the project. The cutting is done with a router equipped with a template guide.

PANTOGRAPHS

Pantographs, like the Black & Decker example in FIG. 11-44, allow the router to duplicate just about any design, letter, or number that you can draw or cut from a magazine or newspaper. Reductions of 40, 50, and 60 percent are

11-44 The bit in the router duplicates the lines that are followed by the stylus. V-groove or small diameter straight bits should be used, at least to start with.

possible. A good degree of operator control is required with units like this because the results depend on how you guide the stylus along the lines of the pattern. If you acquire an accessory of this nature, do some practicing by making shallow cuts with a V-groove bit or a small diameter straight bit. Move the stylus slowly, not quickly, so you can accurately follow the lines you wish to reproduce.

This particular unit can be used with any router having a 6-inch-diameter, or smaller, base. It comes with 40 sheets of patterns for letters, numbers, and a variety of figures and designs. The accessory must be mounted to a sheet of ¾-inch plywood or directly to a bench. The plywood mounting is more practical because you will be able to save bench space by storing the unit when it is not in service.

INLAY WORK

The most common type of inlay work calls for forming specially shaped recesses in a surface to accommodate ready-made practical or decorative veneer assemblies (FIG. 11-45). These intriguing products are composed of dozens of pieces of contrasting species of wood and colors and run the gamut from border strips to complete scenics. Examples of various types of inlays include fraternal emblems, signs of the zodiac, chess and card symbols, birds, bees, and flowers, and purely decorative motifs that can add an artistic touch to any project. Such assemblies are readily available from supply houses like Constantine's in New York.

Many of the assemblies are held together with tape on the face side, the side that should be "up" when the units are glued in place. There are two methods you can use to form the necessary recess. The first method calls for

11-45 Inlay borders are available for work that ranges from the restoration of antiques to modern furniture projects. They are usually sold in 36-inch lengths and in various widths. There is much variety in wood species and color.

making a template by using the inlay as a pattern and cutting the work by using a template guide. The second method calls for marking the pattern then doing the recessing freehand. The latter method is faster but calls for very careful handling of the router. The best procedure is to incise the outline with a sharp knife, using the inlay as a template. Then, with a medium-size straight bit, say ¼ inch, and with depth of cut set to match the thickness of the inlay, clean out the waste between border lines. If the inlay has sharp corners, the job will have to be finished with a knife or, as many experts do, with a very small hand tool called a *router plane*. If you have many similar pieces to install, it's probably wise to take the time to make a special template. The recess should have clean lines and a smooth, level bottom (FIG. 11-46). If you are installing inlay borders or good-size pieces that are square or rectangular, make the border cuts by working with tack-nailed or clamped straightedges.

After the inlay is in place and the glue has dried (special glues are available), remove the paper by moistening it with a damp cloth. You'll find if you wait a minute or two, the paper peels off easily. Sand and finish when you're sure that the inlay and adjacent areas are dry.

Another type of inlay work is shown in FIG. 11-47. Here, homemade strips of material are installed in grooves formed with the router. The joints

11-46 The secret to professional inlay work is all in the recess. Using a sharp knife to outline the inlay's shape is a good way to start.

11-47 Inlay strips can simply be pieces of contrasting material. Joints will be perfect if you form grooves for crossing strips after the first pieces have been installed.

where strips cross will be perfect if you form the grooves for crossing pieces after the first strips have been glued down. The system was used to make the top for the chess table shown in FIG. 11-48. Squares were formed by using the router to make parallel grooves that were filled with strips of contrasting material. Then a second set of parallel grooves running at right angles to the first ones were cut and filled. First form a test groove in some scrap material to judge how to size the inlay strips. When the work is done carefully, the intersecting pieces will look like a preassembled inlay (FIG. 11-49).

11-48 The chess area of the table was made by forming grooves in a plywood panel and then filling them with contrasting strips.

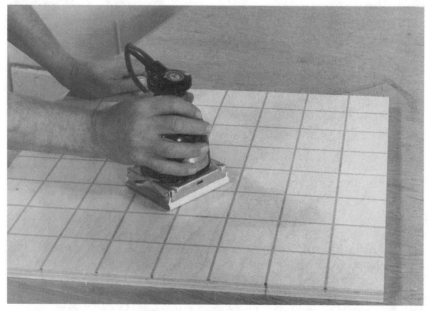

11-49 Every joint is perfect when the second set of grooves is formed after the first inlay strips have been glued down. "White" squares were covered with shellac before the "black" squares were stained.

234   More router applications

INLAY KIT

Something new for router users is an *inlay kit* that is now being offered by quite a few suppliers. This kit consists of a template guide, a bushing that fits over the sleeve of the guide, and a special bit whose diameter equals the wall thickness of the bushing (FIG. 11-50). It can be used to outline the recess for the inlay *and* to form the inlay itself. The only restriction with the kit is that you can't turn a sharp corner because of the diameter of the bushing. Any turn you plan in the design must be an arc whose radius is one half the bushing's diameter. However, it's a minor factor that doesn't overly limit creativity. The product is used by following the steps that are diagramed in FIG. 11-51 and detailed below.

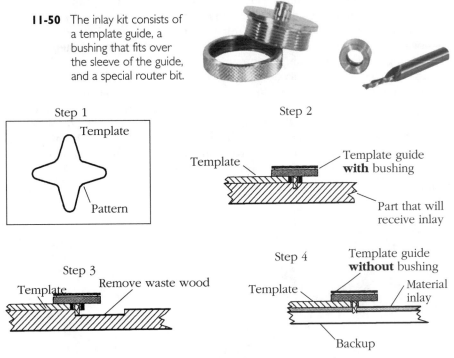

11-50 The inlay kit consists of a template guide, a bushing that fits over the sleeve of the guide, and a special router bit.

Step 1

Template

Pattern

Step 2

Template

Template guide **with** bushing

Part that will receive inlay

Step 3

Template   Remove waste wood

Step 4

Template

Template guide **without** bushing

Material inlay

Backup

11-51 This is how the inlay kit is used. The kit provides for forming a recess plus the inlay that fits precisely.

Step 1. Use ¼-inch hardboard or a similar material to produce the template. It's a good idea to make the template large enough to supply support for the router. Just remember the turn radii restriction.

Step 2. Use clamps, if feasible, or double-face tape to attach the template to the work that will receive the inlay. Work with the template guide *and* the bushing, run the router around the edges of the pattern. This establishes a groove that is the shape of the pattern.

Step 3. Clean out the waste material. If, perchance, you are working with a veneered or laminate surface, you can probably lift or remove the waste with a chisel, or by using a heat gun to soften the adhesive.

Step 4. Use double-face tape or whatever means is suitable to attach the same template to the material that will be the inlay piece and to a backing. Work with the template guide but *without* the bushing to run the router around the edges of the pattern. The result is an inlay that fits precisely the recess that was formed in Steps 2 and 3.

MORTISING FOR HINGES

Mortises required for door hinges are recesses so that they match the thickness, width, and length of one of the hinge leaves. How you go about accomplishing the cuts depends primarily on your frequency of use. Professionals, or amateurs who might be building a house or doing extensive remodeling work, buy or rent equipment that facilitates the work and guarantees accuracy from door to door. An example of available hinge mortising equipment is the set of templates produced by Porter Cable and shown in FIG. 11-52. These templates are adjustable longitudinally along an extrusion that is secured to the edge of the door. Once the templates are positioned for the number of hinges and the spacing between them, the hinge butt template can form accurate hinge mortises on any number of doors. Also, the same setup is used to form the compatible recesses that are required on door jambs. Individual butt hinge templates, in 3½-inch and 4-inch sizes can be purchased.

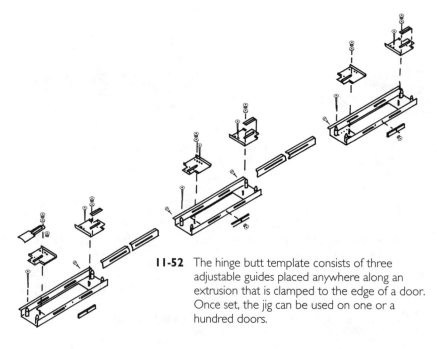

11-52 The hinge butt template consists of three adjustable guides placed anywhere along an extrusion that is clamped to the edge of a door. Once set, the jig can be used on one or a hundred doors.

If this phase of router use is encountered too infrequently enough to justify purchasing or even renting a hinge butt template, there are other acceptable solutions. One of them is simply to use the hinge as a pattern to mark the edge of the door and then make the recess by using the router freehand. For better results, you can make a template guide of the hinge to eliminate the possibility of human error. The template can be tack-nailed to the work (FIG. 11-53) or can be attached to a strip so that clamps can be used to hold the template in place (FIG. 11-54). Results can be just as good as those achieved by other means, but working with homemade templates does make you responsible for the distance between mortises.

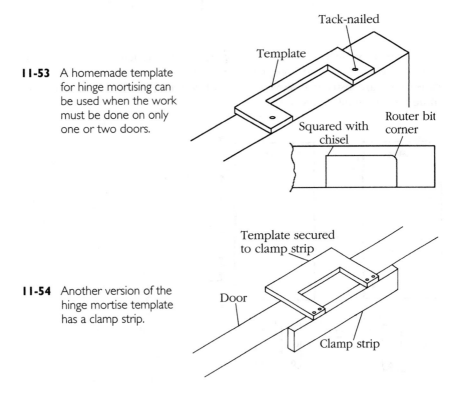

11-53 A homemade template for hinge mortising can be used when the work must be done on only one or two doors.

11-54 Another version of the hinge mortise template has a clamp strip.

Cutting is done with a straight bit that leaves rounded corners. Some hand work with a knife or chisel is required if the hinges have square corners; however, if you wish to avoid such hand work, round corners are available. If you seek out the round-cornered hinges, you'll be ready for hinge installation as soon as routing is complete.

Another solution, if you must square the corners of the mortise is to acquire a "corner chisel" like the one that is sketched in FIG. 11-55. Place the self-aligning tool up against each corner of the recess and then hammer the chisel to incise a perfect 90-degree corner. Since the depth of the cut of the tool can be as much as ⅜ inch, it can be used for more than just shallow hinge mortises.

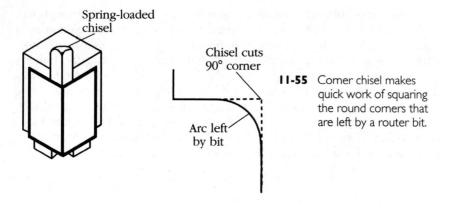

Spring-loaded chisel

Chisel cuts 90° corner

Arc left by bit

11-55 Corner chisel makes quick work of squaring the round corners that are left by a router bit.

Figure 11-56 illustrates some facts pertaining to efficient door hanging. Although two hinges are acceptable on interior doors, which are light because they are usually hollow core, it's becoming more common to use three hinges on all doors.

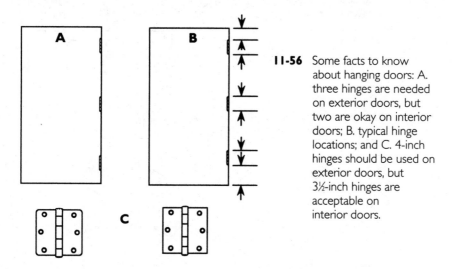

11-56 Some facts to know about hanging doors: A. three hinges are needed on exterior doors, but two are okay on interior doors; B. typical hinge locations; and C. 4-inch hinges should be used on exterior doors, but 3½-inch hinges are acceptable on interior doors.

PANEL DECORATING

Panel decorating refers to the decorative grooving that is commonly seen on cabinet doors, cabinet sides, and drawer fronts (FIG. 11-57). There are many different types of accessories available for the router that make this kind of work easy to do. These accessories are expensive though. The average router user, who is not involved in this activity to a great extent, can get by economically and efficiently by making a special template for the job at hand (FIG. 11-58). The template can be a one-piece unit or an assembly of straightedges, quarter-round parts, and other pieces that have concave or convex arcs. The latter idea, which forms a template when the various pieces are clamped or tack-nailed to the work, was demonstrated in chap-

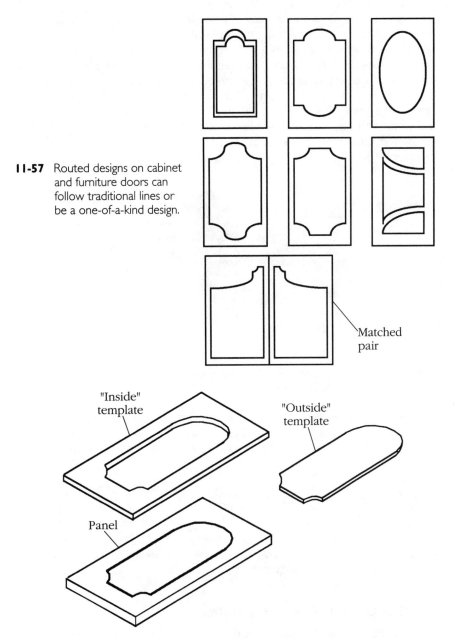

11-57 Routed designs on cabinet and furniture doors can follow traditional lines or be a one-of-a-kind design.

Matched pair

"Inside" template

"Outside" template

Panel

11-58 A conventional template that you make can be used to guide the router.

ter 8, Working with template guides. Working with an assortment of shapes makes the template variable in size and in the groove-shapes that can be produced. When the situation doesn't permit clamping, a disadvantage is that the pieces must be secured by spot-gluing or tack-nailing, and these might not be acceptable.

The advantages of commercial accessories are that they are designed to accommodate panels of various size. Built-in clamps or holders of some sort secure the work and a variety of corner templates that just snap into position are supplied. The Sears Door and Panel Decorating kit (FIG. 11-59) might be of interest to amateurs, or professionals for that matter, because it can be used on workpieces up to 36 × 36 inches and it's in the $80.00 price range. The guide rails are extruded anodized aluminum that are secured to the work with eight clamps. Assuming that all the parts you need to work on are the same size, the assembly needs no further adjustment after it has been organized for the first piece of work.

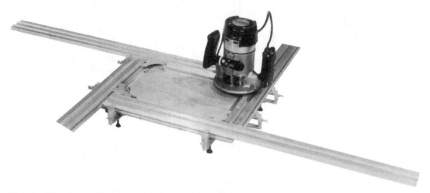

11-59 The Sears Door and Panel Decorating Kit consists of aluminum straightedges and corner templates that just snap into place.

Seven four-piece sets of corner templates (FIG. 11-60) and an extra "blank" set that you can shape for a particular design are part of the kit. An adjustable radius arm is also provided so the tool can be used for forming arcs.

11-60. Corner templates come in sets of four. Seven designs are provided in the Sears Kit, in addition to a blank set that you can use for an original design.

WORKING WITH PLASTIC LAMINATES

Plastic laminates are as common in the modern home as an automobile on the driveway or a TV set in the family room. There was a time when the material was accepted merely as a long-lived covering for a kitchen countertop, but it is now found on furniture and cabinets, on walls and shelves, and even in moisture areas like stall showers. The reasons for increased usage include improvements in the material, better installation methods that are not beyond the scope of amateurs, and the ever-increasing variety of colors and decorative patterns and finishes. Grain patterns and tones of many wood species, butcher block effects, solid colors, slate and marble simulations that can fool the eye, the choice of a high-gloss, suede, satin, or textured finish all contribute to plastic laminate excitement. One of the newest innovations is a solid-color product that has no dark substrate. The joints on edges and corners are almost invisible and can be surface decorated like any solid material (FIG. 11-61).

11-61 A recently introduced plastic laminate has solid color throughout. This eliminates dark lines in joints and makes it possible to surface decorate the material. Current trade names are Wilsonart's "Solicor" and Formica's "Colorcore."

If you see the sheets of plastic laminate as a veneer that is adhered to a core material, you can accept installation as a three-phase operation. The sheet is first cut to approximate size, then it is glued down by following a certain procedure, and finally it is trimmed to match the size and shape of the substrate. The trimming is the operation that requires some finesse, but because it can be done with a router that is equipped with the kind of bits shown in FIG. 11-62, it's a chore that can be accomplished by anyone, with professional results just about guaranteed. The secret of the bits is that the pilot, whether it is ball-bearing or integral, must match the diameter of the bit's cutting circle. As long as you keep the router level and the pilot in contact with the bearing surface, you can't go wrong. The trimmed edge will be perfectly flush with the adjacent surface (FIGS. 11-63 and 11-64).

In step-by-step fashion, the overall procedure is as follows: The surface and edges of the core material (particleboard or plywood) must be smooth and clean. Fill any voids in the core edges before going further. Because most applications call for the surface laminate to cover the top edges of side pieces, edges must be covered first. Cut strips about ¼ inch wider than the edge they will cover. Attach with glue. Because conventional liquid glues require much clamping and a lot of time to dry, it's become standard prac-

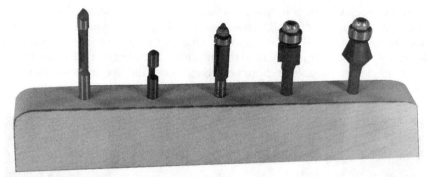

11-62 Types of router bits that are commonly used on plastic laminates. The one on the right leaves a beveled edge; at the left is a plunging design. The bits are solid carbide or have tungsten carbide blades.

11-63 Bits used for trimming do an excellent job because the bearing has exactly the same diameter as the bit's cutting circle.

11-64 Perfect joints are assured if you keep the router level and the bearing constantly against the work edge. Some professionals coat the bearing edge with wax or petroleum jelly as a guard against marring. This is probably needed more when a bit with an integral rather than a ball-bearing pilot is used.

tice to work with a contact cement. The only disadvantage is that the cement bonds on contact, so care must be taken to register mating pieces correctly before you press them together. Read the directions on the contact cement container before using. The length of time it takes for the cement to touch-dry, which it has to do before you join parts, can vary. The instructions will suggest a test, a common one being to touch a cemented surface with a piece of wrapping paper. If the paper doesn't stick, the cement is dry enough to work.

Place the edge pieces so the excess projects above the surface of the core. Then trim the top edge flush with the surface of the core by working with a router or a laminate trimmer (FIG. 11-65). The next step is to cut the surface cover so it is about ¼ inch greater on all edges than the substrate.

11-65 Edges of core material are covered first. The router, in this case a laminate trimmer, cuts the strip so it is flush with the core's surface.

Formica Corporation

After applying the contact cement and waiting for it to dry, place the cover in position but with strips of wood or dowels between it and the core (FIG. 11-66). This ensures correct registration before the cover is pressed into place. Remove the separation pieces as you go along. Be sure to apply pressure over the entire area, either by tapping with a hammer and a length of hardwood or by using a roller. An idea that works is to use an ordinary rolling pin. The last step is to work with the router to trim off excess material (FIG. 11-67).

One of the more tricky aspects of plastic laminate cutting comes when it is necessary to join pieces end to end or when the covering must make a

Formica Corporation

11-66 Remove the strips, in this case dowels, as the cover is pressed down. When full contact is made, immediately apply pressure over the entire area. Do this with a hammer and a strip of hardwood, or something like a rolling pin.

Formica Corporation

11-67 The last step is to trim the laminate so it will be flush to the edges.

90-degree turn. The mating edges of the joint must be perfect in order for the project to appear seamless. One system you can use for perfect butt joints is shown in FIG. 11-68. The two parts are elevated on boards or pieces of plywood and placed so the edges to be joined are separated by less than the diameter of the straight bit that will do the cutting. The router is guided by a straightedge that is secured parallel to the cut line. Another way would be to overlap the edges and guide the router through the cut by clamping a straightedge on each side of the cut line as tracks for the tool.

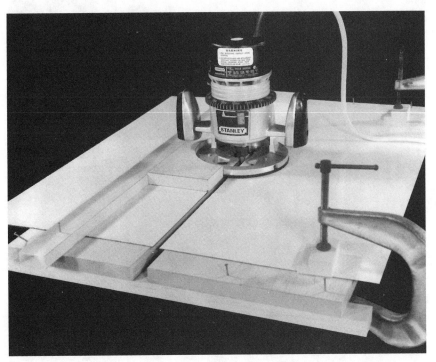

11-68 You'll get perfect butt joints if the mating edges of the pieces are trimmed at the same time.

Miter joints can be cut in similar fashion (FIG. 11-69). You must be sure to place the pieces so the cut angle will be exact. In this case, the mating pieces are placed together face side to face side. When the cutting is carefully done, the joint will be tight and invisible (FIG. 11-70).

Figure 11-71 shows examples of how you can work with plastic laminates as if they were wood veneers. Often, scrap pieces can be utilized this way on projects like small tables, boxes, clock faces, and so on.

THE ROUTER AS A THREADING TOOL

Cutting threads in wood may seem odd, but the technique has many practical applications (FIGS. 11-72 thru 11-74). There is nothing wrong with getting

11-69 Cut mating edges together for miter joints. In this case the parts to be joined are held together face side to face side. Be sure the pieces are clamped in place at the correct angle.

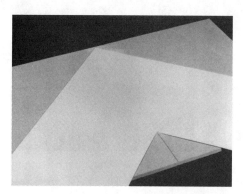

11-70 Careful cutting results in tight joints that are hard to detect.

11-71 Plastic laminates can be worked almost like wood veneers so interesting inlay effects are possible.

11-72 You can form wooden nuts and bolts when the router is organized for threading operations.

11-73 Other projects include light-duty, conventional, or special application clamps.

11-74 My book binding jig uses threaded dowel to apply pressure.

into some areas of shop work just for the fun of it. Often adding a threaded post or securing an assembly with nuts and bolts made of wood can provide eye-catching details to a project.

This type of work is usually done with special thread boxes that are used by hand, but a recent innovation allows quick and precise thread forming in wood, and even in some plastics, by using a router. The accessories required (FIG. 11-75) are offered by The Beall Tool Company and can be used with any router. The parts can be purchased individually or as a complete kit that includes threading equipment (for outside threads) and taps (for inside threads) in sizes of ½, ¾, and 1 inch.

The threader, which serves as a base for the router (FIG. 11-76), is bolted to a block of wood that is then secured in a vise or clamped to a benchtop. Special dies are used in the base so that after the cut is started, feeding the

11-75 The Beall Tool Company threading tools can be used with a router and can be purchased as the set shown here or as separate components.

11-76 The router mounts on a special base that is secured to a block of wood.

work through at the correct speed is automatic. The router bit supplied is an HSS double-ended, three-flute spiral type that was specially designed for the purpose of forming threads.

The only crucial adjustment that is required of the worker is the projection of the bit. A little trial and error is required before the threads are perfect, but once the cut depth is established, you can turn out threaded pieces by the yard.

The dowels to be threaded should be selected with care. Reject any that are undersize or more oval than round. Dowels that are slightly oversize can be made perfect by driving them through the sizing plate. Internal threads are formed in fairly routine fashion with the taps. The taps that are supplied were specially designed with long pilots to keep them vertical in the hole and assure perpendicular threads.

CHAIN-MAKING TECHNIQUE

Using the router to form links for wooden chains might not be of interest to all router users, but the techniques involved are very intriguing and might come in handy for small project components requiring special setups, or when making wooden rings. Creating links for wooden chains is another activity that demonstrates the flexibility of the router and worth knowing, even if rarely used. The procedure is explained as follows:

Step 1. Prepare the number of pieces required for the links by sawing. Make a cardboard template to mark the outside and inside shape of the links on each piece. Shape the outside with a handsaw, a jigsaw, or band saw. Sand the outside edges smooth. Then, remove the bulk of the waste from the interior area by boring overlapping holes (FIG. 11-77).

11-77 Blanks for the wooden links are cut to size and shaped to outside form. Remove the bulk of inside waste by boring overlapping holes.

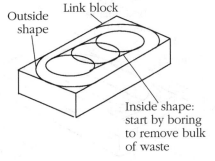

Outside shape

Link block

Inside shape: start by boring to remove bulk of waste

Step 2. Make the holding fixture shown in FIG. 11-78. The frame provides a tight fit for the link block. The opening through the base, which serves as a template, is sized to suit the inside shape of the link. Use a straight, piloted router bit to finish shaping the inside of the link.

Step 3. Use the same holding fixture from Step 2 to hold the link while rounding over the inside, top, and bottom edges (FIG. 11-79).

Step 4. Make a holding fixture that exactly suits the inside shape of the links (FIG. 11-80). Use the rounding-over bit to shape the outside edges of the links.

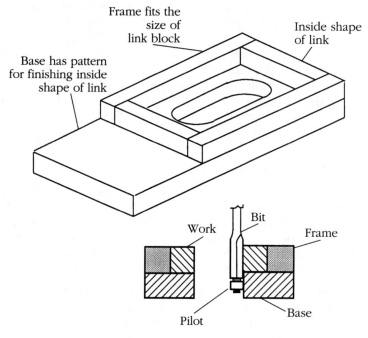

11-78 In step 2, a box frame is made that serves as a jig so the inside of the links can be routed to final form. A straight, piloted bit is used.

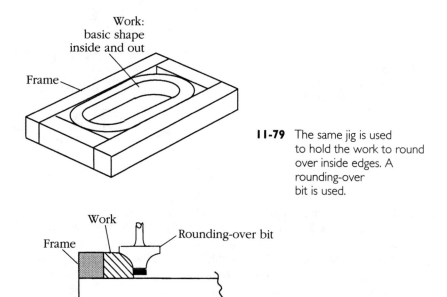

11-79 The same jig is used to hold the work to round over inside edges. A rounding-over bit is used.

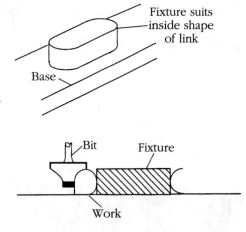

11-80 For the final step, create another fixture with a plug that suits the inside shape of the link. The rounding-over bit is used to shape the perimeter of the links.

A factor that is not shown in the illustrations is that the piloted bit must have some bearing surface. This means that the inside and outside edges of the links will have a *flat* area. You can either live with this or do some sanding to get rid of it. Half of the links will have to be split on the centerline before all links can be connected (FIG. 11-81). You can do this with a knife or a thin-bladed saw. A coping saw with a very fine blade will do if nothing else is available. Reassemble the parted links after they have been inserted through full links by coating mating edges with glue and holding them together with small clamps or heavy rubber bands.

11-81 Alternate links must be split on a centerline so they can interlock with full links.

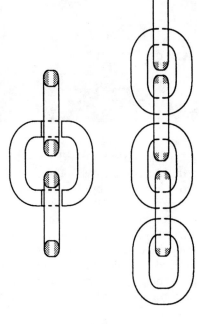

THE ROUTER AS A LATHE ACCESSORY

If you own a lathe, you can view the portable router almost as if it were a motorized lathe chisel. The router is suitable for such use because home-made jigs allow it to be moved longitudinally on the lathe's working axis to form reeds and flutes or held still for circumferential cutting. If you recall the fluting jig described in chapter 10, you'll recognize the similarity be-tween that jig and the lathe arrangement shown in FIG. 11-82. The jig for the lathe can be designed with raised sides that serve as tracks for the router or with separated support areas to hold the router level while it moves accu-rately using an edge guide (FIGS. 11-83 and 11-84).

11-82 It's easy to do fluting or reeding on a lathe-mounted spindle when a trough-type jig is installed. With this jig, the router can be guided longitudinally over the centerline of the work.

Most lathes have indexing devices to maintain spacing between longi-tudinal cuts (FIG. 11-85). If the lathe does not have this feature, you can still work accurately by using the strip of paper system that was described in chapter 10, Special jigs you can make.

For cuts on the circumferences of spindles, you can hold the router in a fixed position by using clamps or by bracing it against a stop that is fixed to the jig (FIG. 11-86). The cut depth can be controlled by adjusting the bit's projection while both spindle and router bit are turning or by adjusting bit projection beforehand and then sliding the router to make contact. There is opportunity here for experimentation. One example is shown in FIG. 11-87.

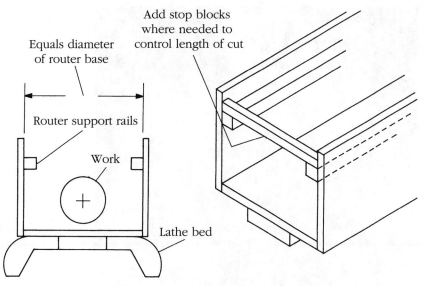

Add stop blocks
where needed to
control length of cut

Equals diameter
of router base

Router support rails

Work

Lathe bed

11-83 This type of lathe jig provides tracks for the router. It must be made so the bit doing the cutting and the workpiece have the same vertical centerline. Stops placed across the side members of the jig are used when you wish to limit the length of the cuts.

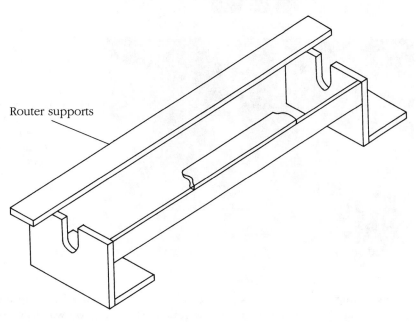

Router supports

11-84 Another design for a lathe jig you can make uses the router like a motorized lathe chisel. In this case the jig supplies support areas. You can control the cutting path of the router by using an edge guide.

11-85 The router is equipped with a special subbase so it easily spans the open area between support surfaces. Notice that the auxiliary base is a $\frac{1}{16}$-inch-thick piece of aluminum. This is not crucial for the application, but it does demonstrate that various materials can be used for subbases that you manufacture.

11-86 Peripheral shaping is done with the router in a fixed position. Both the router bit and the workpiece are turning. Cuts like this can't be made with piloted bits.

In this example, the spindle is formed to a particular shape by using the template to guide the movements of the router. This is a good way to work when you need many similar pieces. Fluting and reeding cuts are made with the work in a fixed position and only the router working. Peripheral cuts are made with both the lathe and the router turning.

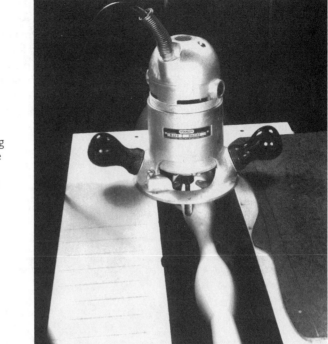

11-87 Contours on spindles can be shaped by using templates to guide the router. Work with rotary rasps and burrs instead of regular router bits.

THE ROUTER AS A MOTORIZED LATHE CHISEL

Unite a portable router with the jig on display in FIG. 11-88 and you can do some surprising, mechanized lathe work. The advantages of the arrangement are not difficult to imagine. You can easily duplicate shapes—utilizing the speeds of both router and lathe results in cuts that require little sanding; form specific shapes that can be duplicated; and work freehand for wide coves and such.

11-88 Making a jig like this turns the portable router into a motorized lathe chisel.

The jig, which is detailed in FIGS. 11-89 and 11-90, may need some modifying to suit the tools you will use. The carriage assembly, for example, should accommodate your router. Here, it's best to make the assembly as narrow as you can so swivel action will be limited as little as possible. You may have to change the shapes of side pieces to provide access to the router's depth-of-cut lock. Size the T-shaped guide to suit the cross-sectional dimensions of the lathe bed. Other components should work pretty much as dimensioned. Be sure the slide fits the slide base precisely. It should move with little pressure and with zero side play.

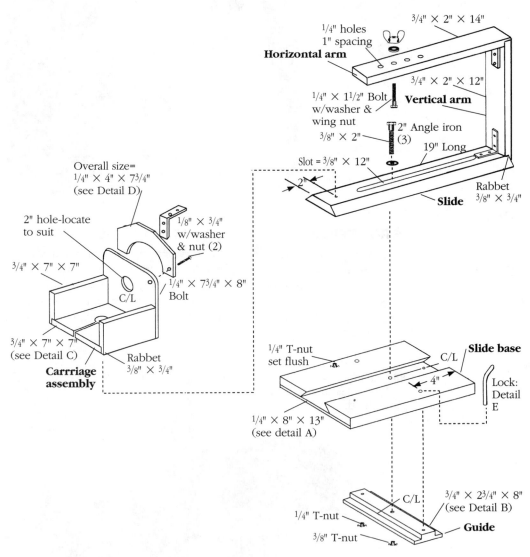

11-89 Construction details for the lathe jig. The vertical and top horizontal members add stability and provide a means for securing the chip shield.

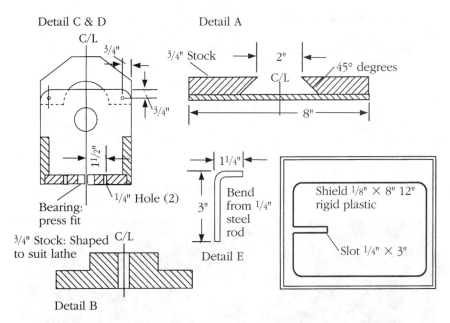

Detail C & D

C/L

3/4"

Bearing:
press fit

1/4" Hole (2)

1 1/2"

3/4" Stock: Shaped
to suit lathe

Detail B

C/L

Detail A

3/4" Stock

2"

C/L

45° degrees

8"

1 1/4"

3"

Bend
from 1/4"
steel
rod

Detail E

Shield 1/8" × 8" 12"
rigid plastic

Slot 1/4" × 3"

11-90 Construction details for the lathe jig shield. The shield can be situated at right angles to or parallel with the lathe bed. The shield can't be used on some applications so safety goggles must be used.

11-91 The jig has several locking devices. The L-shaped device prevents lateral motion. The bolt at the rear is tightened when it's necessary to keep the slide base in a fixed position.

The locking devices, shown in FIG. 11-91 on previous page, are used for various functions. Most times, the main lock, which is the ⅜-inch × 2-inch bolt that passes through the slide base and threads into the T-nut that is in the guide, is tightened and then loosened just a bit so the jig can move along the ways. The L-shaped lock prevents lateral motion when the slide-mounted carriage that bears the router is moved directly forward for circumferential cuts. Drill the holes for the lock slightly undersize so the unit will fit tightly.

For longitudinal cuts, like those required to shape cylinders, bring the router/carriage unit forward for the cut depth required and after locking its position, move it slowly from right to left (FIG. 11-92). How deep you can cut will depend on the router's power; remember, overdoing is never wise. Making several passes usually results in smoother work.

11-92 Cylinder-type work is accomplished by moving the entire jig parallel to the lathe centerline. The main lock is tightened, then loosened just enough to allow the jig to move. Keep the lathe bed and the jig's contact areas waxed.

Fluting cuts and other like operations are done in similar fashion except that the lathe's indexing device is used to space the cuts and to keep the work in proper position for each of them (FIG. 11-93).

For circumferential cuts (FIG. 11-94), the slide base and guide are locked in a fixed position and the router/carriage unit is moved forward. A plunge router makes it easy to control the depth of the cuts when the shapes must be similar.

For freehand work, the carriage is loosened just enough to allow a swivel action (FIG. 11-95). There are limitations with this procedure since the width of the carriage permits just so much rotation. Don't try to do more by "extending" the router bit in the collet. A bit that isn't seated correctly can loosen and do harm.

11-93 The setup for fluting is the same as for cylinder forming.

Some amount of faceplate work is possible, but here too, the width of the carriage limits applications. Note, in FIG. 11-96, that the cutter is positioned on the "up" side of the turning. This, so the rotation of the bit will be *against* the rotation of the work; a factor that opposes the rule to follow when using lathe chisels.

Whatever operation you perform, follow the rules that apply to correct speed for both router and lathe.

11-94 Circumferential cuts are formed by moving the router/carriage unit directly forward.

11-95 The router/carriage swivel action is used to create cove-type cuts. There is a limit to how much you can do because of the width of the carriage.

11-96 The jig can be positioned for some amount of faceplate turning. Cutting is done on the "up" side of the disc so the bit's rotation will be against that of the disc. Bits used with the jig must be of the pilotless type.

12

The router
as a shaper

The portable router and the stationary shaper are two different tool concepts. They have enough in common, however, that a router secured in an inverted position in a special stand is so much a counterpart of the other that the differences in appearance and use are minimal (FIGS. 12-1 and 12-2).

12-1 Anyone interested in making a router/shaper stand should think in terms of imitating the features of a good-size, regular shaper table, like this Rockwell (now Delta) industrial tool.

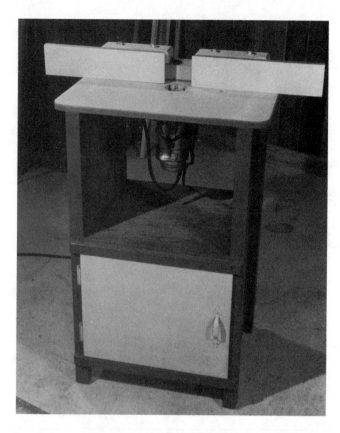

12-2 The homemade stand should be a counterpart of the commercial version. It lacks, among other things, a slot for a miter gauge.

Too often, though, the router is organized for stationary shaper work by mounting it under a slab of plywood that has a hole through it for the router bit. A clamped-on strip of wood serves as a fence. Working this way handicaps anyone who wants to thoroughly utilize this setup. For one thing, the fence should consist of individually adjustable *infeed* and *outfeed* sections. Then, when the cut removes the entire edge of the stock, for example, the outfeed fence can be brought forward to provide support for the work *after* it has passed the cutter (FIG. 12-3). Another factor is that edges on curved pieces can't be shaped by working against a fence. A regular shaper provides *fulcrum pins* for this facet of shaping, and so should a homemade router-using version (FIGS. 12-4 and 12-5). The function of the pins is to provide a bracing point for the work so it can be held firmly while it is advanced slowly to make contact with the cutter. Freehand work on a shaper is done with the workpiece bearing against collars that are mounted on the spindle along with the cutter (FIG. 12-6). When a router is used, the bearing surface for the work must be provided by a pilot on the router bit (FIG. 12-7).

APPLICATIONS

There isn't much you can't do with a router/shaper setup. By working with conventional router bits, you can form decorative edges on straight, round,

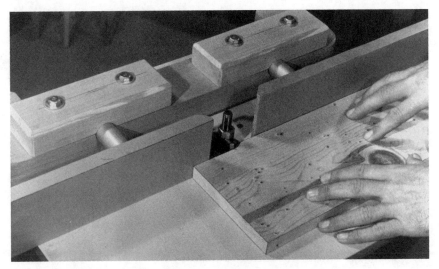

12-3 Infeed and outfeed fences should be individually adjustable.

12-4 A regular shaper provides for the installation of fulcrum pins that are needed when doing freehand shaping. The arrow indicates feed direction, which is always against the cutter's direction of rotation.

or curved workpieces; do rabbeting and dadoing; joint edges; cut grooves; form tenons; and more. Even more exciting is that you can work with the new types of bits being produced—bits designed for practical purposes, but which can't be used in traditional router fashion. (They can only be used when the router is mounted in a stand.) Most of the newcomers make it possible for router users who add a shaper table to do work like panel raising and forming those intriguing mating configurations like cope and stile cuts that are required for frame and panel constructions. Cabinet doors (FIG. 12-8) are now included in the list of projects you can accomplish with a portable router.

12-5 A homemade version of a shaper should also provide for fulcrum pins, one on the infeed side and a second one on the outfeed side.

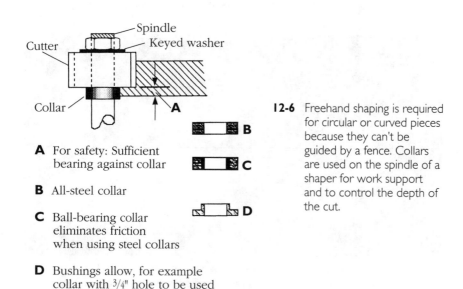

Spindle
Cutter
Keyed washer
Collar
A

A For safety: Sufficient bearing against collar

B All-steel collar

B

C

C Ball-bearing collar eliminates friction when using steel collars

D

D Bushings allow, for example collar with 3/4" hole to be used on 1/2" spindle

12-6 Freehand shaping is required for circular or curved pieces because they can't be guided by a fence. Collars are used on the spindle of a shaper for work support and to control the depth of the cut.

Bits in this category may be purchased individually or in sets. Some include extra bits like the door lip and the glue joint cutter that are included in the Freud set shown in FIG. 12-9. The glue joint cutter is handy when the door panel is made of solid wood rather than a material like plywood. One reason why you might opt to glue solid wood together to form a door may be because available stock might not be wide enough to suit the width of the door. Another reason is that narrow pieces, joined edge-to-edge, are less likely to warp than a one-piece board. It often happens that a board is concave (cupped) across its width. A common method used to eliminate the distortion is to saw the board into three pieces and then reassemble with the

12-7 When freehand shaping is done on a router/shaper stand, work support and cut depth are provided by the pilot on the cutter.

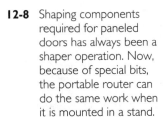 **12-8** Shaping components required for paneled doors has always been a shaper operation. Now, because of special bits, the portable router can do the same work when it is mounted in a stand.

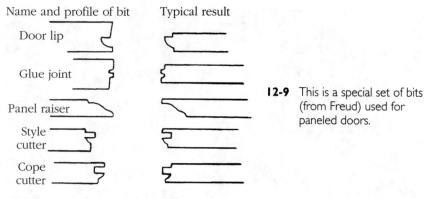

Name and profile of bit   Typical result

Door lip

Glue joint

12-9 This is a special set of bits (from Freud) used for paneled doors.

Panel raiser

Style cutter

Cope cutter

Stile and cope cutters are a matched set for shaping edges of stiles and rails

center piece inverted. Other ideas that are used to create flat, solid wood panels are shown in FIG. 12-10. The glue joint cutter is a big help when you are involved in this kind of work.

The door lip cutter shown in FIGS. 12-11 and 12-12 is used to form the edge that is often seen on kitchen cabinet doors and similar projects. Some variation in the height of the shoulders without affecting the shape and size of the quarter-round lip is possible. You can also work so the top shoulder is completely eliminated (FIG. 12-12). The changes from a full profile cut depend on the height of the cutter in relation to the work. Other sets of bits used with a router for paneled door construction are shown in FIGS. 12-13 and 12-14.

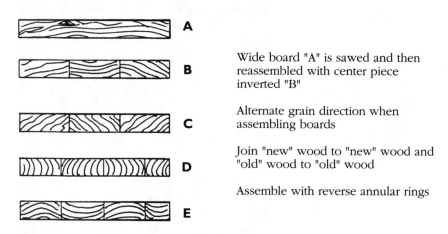

A

B Wide board "A" is sawed and then reassembled with center piece inverted "B"

C Alternate grain direction when assembling boards

D Join "new" wood to "new" wood and "old" wood to "old" wood

E Assemble with reverse annular rings

12-10 Common methods used to guard against warpage and joint separation when door panels are made of solid wood.

The cutters that are used for this phase of router work must not be treated too casually. All of them have ½-inch shanks, which indicates that they be used only with a heavy-duty tool. Even so, it's wise to avoid exces-

12-11 Hinged doors often have edges that are shaped this way. The cut is made in one pass by using a door lip cutter. This type of door is usually used with very simple frames.

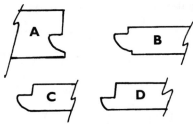

A Profile of door lip cutter
B Cut made to include shoulder
C Cut without shoulder
D Modified quarter round lip
Variations depend on the height of the cutter in relation to the work

12-12 You can vary the shape produced by a door lip cutter by how you adjust the height of the cutter above the table in relation to the work.

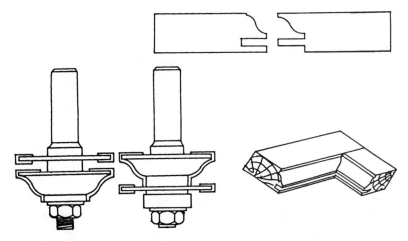

12-13 Special sets of stile and rail cutters are produced by Grissly Imports, Inc.

sively deep cuts and fast-feed rates. The size of the bits alone should inspire respect and prompt you to use them with care and good router judgment.

Another set of bits that should be mentioned because they must be used in a router/shaper setup, is the seven-piece Sears Crown Molding Kit (FIG. 12-15). It contains all the cutters and arbors that are required for producing

Quick view of how the "Door shop" bits are used

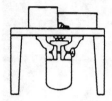

The configuration in the face of rails and stiles is shaped with the **ogee** bit

The **ogee** bit is raised and, after test cuts, is used to form the cope cut on the end of the rails

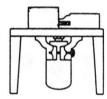

The **slot-cutting** bit forms panel-insert grooves in the stiles and rails

The **slot-cutting** bit is lowered to complete the shape of cope-cut rail-ends

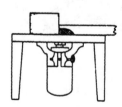

The **panel-raiser** bit shapes the edges of the panel that will be inserted in the frame

12-14 Some of the cutters in the Door Shop set from Zac Products, Inc., do double duty.

12-15 The cutters in the Sear Crown Molding Kit must also be used in a router/shaper setup.

standard or original molding shapes. Chapter 3, Router bits, has more information on the Sears Crown Molding Kit. If you own a Delta light-duty or heavy-duty stationary shaper, you can use it with router bits by replacing the standard spindle with the special router spindle shown in FIG. 12-16.

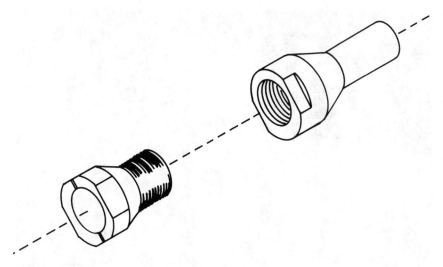

12-16 This Rockwell (Delta) shaper adapter allows the shaper to work with conventional router bits that have ¼-inch or ½-inch shanks.

COMMERCIAL ROUTER/SHAPER TABLES

The Porter Cable product in FIG. 12-17 is a husky unit with a 16-inch × 18-inch nonconductive, "Benelex" table. It has most of the features of a regular shaper including individually adjustable fences that can also be moved longitudinally to minimize the opening around the cutter. Various table inserts provide openings of 1¼, 2, and 3 inches. A feature I like is the built-in, 20-amp, lock-

12-17 Porter Cable router/shaper table can be purchased with or without a heavy-duty 1½-horsepower router.

able key-type, double-pole switch. The router is easily turned on or off without having to reach under the table. The table is available, complete with a 22,000-rpm, heavy-duty, 1½-horsepower router that will accept ¼-inch and ½-inch shank router bits—or comes alone. The table is drilled to accept most Porter Cable routers, but new holes can be drilled by the customer when other brand routers are used.

The new Craftsman (Sears) unit (FIG. 12-18) has a 14-inch × 24-inch die cast aluminum table whose work surface can be doubled by the addition of extra-cost extensions. It features a universal mounting plate that will accommodate most routers and a built-in sawdust collection port that accepts a 2½-inch vacuum hose. The 4-inch-high, polystyrene fence incorporates some novel but practical ideas. An interlocking carriage, actually a pusher/clamp combination, secures workpieces for end cuts that can't be accomplished safely freehand (FIG. 12-19). Most shapers have individually adjustable fences so the *outfeed* fence can be adjusted to support the work when the entire edge of the stock is removed. The Craftsman fence is a single unit but it has an extra *jointing fence* on the outfeed side that can be brought forward as shown in FIG. 12-20. Thus, it can be adjusted to provide necessary support for the work after it passes the cutter.

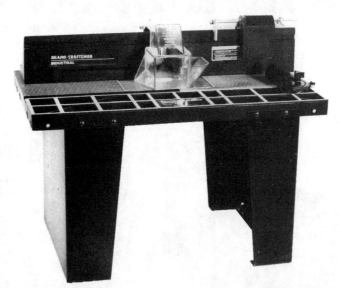

12-18 New Craftsman router/shaper table has some novel but practical ideas for guiding and supporting various types of work. The tool is supplied as shown, plus a miter gauge. An optional steel floor stand is available.

The mounting plate has threaded holes to receive what the manual calls a *starting pin*. This must be inserted whenever shaping is done on curved edges (FIG. 12-21). There will be more information about this type of shaping work later in this chapter.

12-19 A combination pusher and clamp that slides along the fence is designed for safely making end cuts on narrow stock, a chore that should not be attempted freehand.

12-20 Adjustable *jointing fence* is built into the outfeed table. It supplies support after the work has passed the cutter when the cut removes the entire edge of the stock.

12-21 A *starting pin* threads into holes that are in the mounting plate. Work is braced against the pin and slowly advanced until it bears firmly against the pilot on the router bit.

HOMEMADE ROUTER/SHAPER STANDS

There are advantages to making your own router/shaper stand. It can be designed as a floor model to use like any individual tool and can include a cabinet area or drawers for storage of accessory equipment. Table size is optional. A groove for an on-hand miter gauge can be an important factor in the stand's design. Also, you can make attachment arrangements that are exactly right for a router you already own.

I've made several for my own shop, each new one coming closer to an optimum design. Construction details for an early version that served for quite a long time are shown FIG. 12-22. It did not have a slot for a miter gauge, though. I became convinced that even a homemade version should have guards to protect the operator and an electrical arrangement to turn the tool on and off conveniently, without having to use the switch on the tool.

The router/shaper stand that is now standard in my shop has a good-size table and includes all the features of a good shaping machine. Large wood fences are individually adjustable and can also be moved to minimize the opening around the cutter. An adjustable guard is part of the fence assembly that is used for straight line cuts, and there is a slot in the table sized to suit a miter gauge that was already in the shop (FIG. 12-23). The design provides fulcrum pins and a second guard that is used when doing freehand shaping (FIG. 12-24). The on-off switch (FIG. 12-25) is located where it's easy to get at. Having a convenient on-off switch is also better for the router. It's not good for the tool to keep running unnecessarily, which is likely to happen when it's a nuisance to get to a built-in switch.

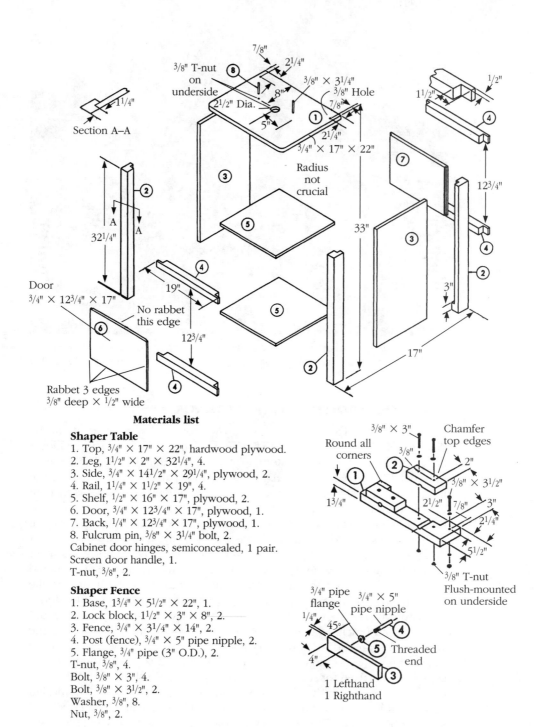

7/8"

2 1/4"

3/8" T-nut
on
underside

3/8" × 3 1/4"

3/8" Hole

8"

7/8"

2 1/2" Dia.

5"

1/2"

1 1/2"

④

1 1/4"

Section A–A

②

③

④

12 3/4"

④

②

A A

32 1/4"

③

⑤

2 1/4"

3/4" × 17" × 22"

Radius
not
crucial

①

33"

⑦

③

Door
3/4" × 12 3/4" × 17"

④

19"

⑤

No rabbet
this edge

12 3/4"

⑥

⑤

②

3"

②

17"

Rabbet 3 edges
3/8" deep × 1/2" wide

④

Materials list

Shaper Table
1. Top, 3/4" × 17" × 22", hardwood plywood.
2. Leg, 1 1/2" × 2" × 32 1/4", 4.
3. Side, 3/4" × 14 1/2" × 29 1/4", plywood, 2.
4. Rail, 1 1/4" × 1 1/2" × 19", 4.
5. Shelf, 1/2" × 16" × 17", plywood, 2.
6. Door, 3/4" × 12 3/4" × 17", plywood, 1.
7. Back, 1/4" × 12 3/4" × 17", plywood, 1.
8. Fulcrum pin, 3/8" × 3 1/4" bolt, 2.
Cabinet door hinges, semiconcealed, 1 pair.
Screen door handle, 1.
T-nut, 3/8", 2.

Shaper Fence
1. Base, 1 3/4" × 5 1/2" × 22", 1.
2. Lock block, 1 1/2" × 3" × 8", 2.
3. Fence, 3/4" × 3 1/4" × 14", 2.
4. Post (fence), 3/4" × 5" pipe nipple, 2.
5. Flange, 3/4" pipe (3" O.D.), 2.
T-nut, 3/8", 4.
Bolt, 3/8" × 3", 4.
Bolt, 3/8" × 3 1/2", 2.
Washer, 3/8", 8.
Nut, 3/8", 2.

3/8" × 3"

Round all
corners

Chamfer
top edges

3/8"

①

②

2"

3/8" × 3 1/2"

13/4"

2 1/2"

7/8"

3"

2 1/4"

5 1/2"

3/8" T-nut
Flush-mounted
on underside

3/4" pipe
flange

3/4" × 5"
pipe nipple

1/4"

45°

④

⑤

Threaded
end

4"

③

1 Lefthand
1 Righthand

12-22 Construction details of my early homemade router/shaper stand.

12-23 My current router/shaper stand, which you can duplicate, evolved from earlier models. It's very similar to the regular shaper. The miter gauge and the attached hold-down are shopsmith items.

12-24 This unit provides for infeed and outfeed fulcrum pins and a guard for freehand shaping.

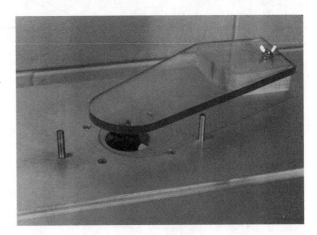

12-25 An external switch, located in an easy-to-reach position, can make the router easier to turn on and off.

Construction details for the new router/shaper cabinet and its components, which include guards to make, are provided in FIGS. 12-26 through 12-29. Table 12-1 provides a list of materials. Unlike some commercial tables, this unit is designed to accommodate any router. The only part of the proj-

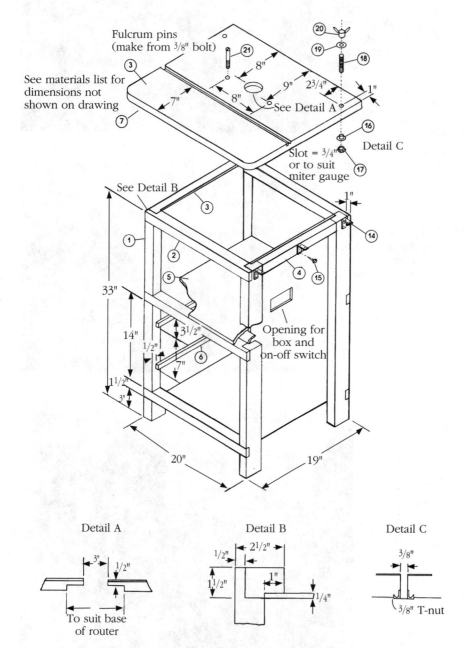

12-26 Construction details for the router/shaper cabinet assembly.

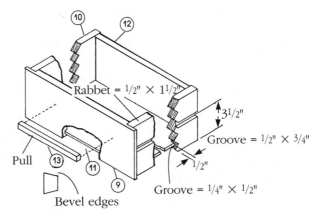

12-27 Construction details for the drawers for the router/shaper cabinet.

Rabbet = $1/2$" $\times$ $11/2$"

Groove = $1/2$" $\times$ $3/4$"

$31/2$"

$1/2$"

Groove = $1/4$" $\times$ $1/2$"

Pull

Bevel edges

ect that needs to be customized is the recess on the underside of the table, which has to be shaped to fit the router's base. Use the subbase as a pattern for the attachment holes. Screws that secure the router are driven from the top. Counterbore or countersink the holes so the screwheads won't project above the table's surface. I chose to install two drawers, but if you wish, you can use the bottom area as a cabinet by installing a hinged door.

Table 12-1 Materials list for the router/shaper cabinet.

Key	Part	#Pieces	Size	Material
1	Leg	4	$1\frac{1}{2}$" $\times$ $2\frac{1}{2}$" $\times$ 33"	Fir
2	Rail	6	$1\frac{1}{2}$" $\times$ $1\frac{1}{2}$" $\times$ 20"	Fir
3	Side	2	$\frac{1}{4}$" $\times$ 16 $\times$ 20"	Hardboard
4	Brace	2	$1\frac{1}{4}$" $\times$ $1\frac{1}{2}$" $\times$ 14"	Fir
5	Shelf	1	$\frac{1}{4}$" $\times$ 17" $\times$ 19"	Hardboard
6	Guides	4	$\frac{1}{2}$" $\times$ $\frac{3}{4}$" $\times$ $18\frac{1}{2}$"	Fir
7	Table	1	$1\frac{1}{8}$" $\times$ 21" $\times$ 24"	(See note)
8	Cover	1	.025" $\times$ 21" $\times$ 24"	(See note)
Drawer				
9	Front	2	$\frac{3}{4}$" $\times$ 7" $\times$ 18"	Pine
10	Side	4	$\frac{3}{4}$" $\times$ 7" $\times$ $17\frac{1}{2}$"	Pine
11	Bottom	2	$\frac{1}{4}$" $\times$ 16" $\times$ $17\frac{1}{2}$"	Hardboard
12	Back	2	$\frac{3}{4}$" $\times$ 6" $\times$ $15\frac{1}{4}$"	Pine
13	Pull	2	$\frac{1}{2}$" $\times$ $\frac{3}{4}$" $\times$ 10"	Pine
Hardware				
14	Connector	6	$\frac{1}{2}$" $\times$ $1\frac{1}{2}$" $\times$ $1\frac{1}{2}$"	Metal angles
15	Screws	24	#8 $\times$ 1"	Roundhead
16		4	$\frac{3}{8}$"	T-nut
17		2	$\frac{3}{8}$"	Nut
18	Stud	2	$\frac{3}{8}$" $\times$ 4"	Threaded rod
19		2	$\frac{3}{8}$"	Fender washer
20		2	$\frac{3}{8}$"	Wing nut
21	Fulcrum pin	2	$\frac{3}{8}$" $\times$ $2\frac{3}{4}$"	Make from bolt

Table 12-1 Continued.

Key	Part	#Pieces	Size	Material
Fence assembly				
22	Base	1	1½" × 5½" × 24"	Fir
23	Guide	2	1⅛" × 1⅛" × 5½"	Fir
24	Fence support	2	1½" × 5" × 10"	Fir
25	Fence	2 (1L/1R)	1" × 3½" × 14"	Fir
26	Guard guide	1	1½" × 3⅛" × 8"	Fir
27	Guard support	1	1½" × 5" × 10"	Fir
28	Shield	1	½" × 5" × 8"	Lexan
Freehand shaping guard				
29	Height block	1	1½" × 5" × 5"	Fir
30	Shield	1	½" × 10¼" × 14½"	Lexan
Hardware				
31		2	¼"	T-nut
32	Stud	2	¼" × 4"	Threaded rod
33		4	¼"	Fender washer
34		2	¼"	Wing nut
35		1	⅜"	T-nut
36	Stud	1	⅜" × 5"	Threaded rod
37		1	⅜"	Fender washer
38		1	⅜"	Wing nut
39		6	#9 × 1½"	Roundhead screw
40		4	½" × ½"	Sq. washer

NOTES:
Table (#7) = made from ready-made glued-up pine slab
Cover (#8) = Wilsonart's aluminum "decorative metal" or do-it-yourself aluminum
Hardwood, like maple or birch, can be substituted for fir

THE ROUTER/SHAPER AT WORK

A difference between a regular shaper and a router/shaper setup is that a shaper might have a reversible direction of rotation while a router turns only in one direction. This means that a shaper operator can choose to move work from left to right or right to left. On a router/shaper, work feed direction is from right to left. Cutting is always done by moving the work *against* the cutter's direction or rotation (FIG. 12-30).

Figure 12-31 supplies important information concerning how fences should be positioned in relation to the cut being made. The bearing surface of both fences are on the same plane when only part of the work edge is removed. When the entire edge of the stock is removed, as in a jointing cut, the outfeed fence is brought forward a distance that is equal to the depth of the cut. If this isn't done, the work will have no support after it has passed the cutter. In addition to being set for the cut, the fences should be adjusted longitudinally to minimize the open area around the cutter.

It's always a good idea, when cutting across end grain, to use a miter gauge to advance the work (FIG. 12-32). There are no options when the work

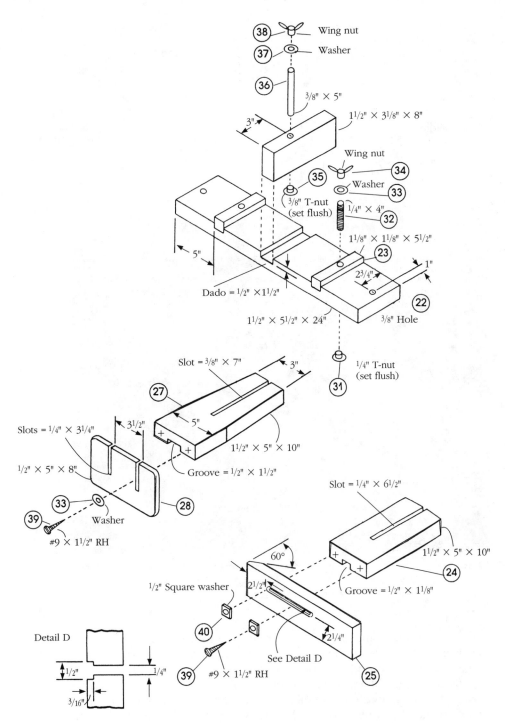

12-28 Construction details for the fence assembly and the adjustable guard that is used for straight line cuts.

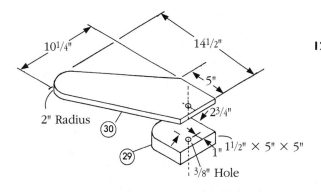

10¼" 14½"

5"

2" Radius 2³/₄"

(30)

(29) 1" 1½" × 5" × 5"

3/8" Hole

12-29 Construction details for the guard used for freehand shaping. Note that the stud that secures the fence base is also used for the freehand shaping guard.

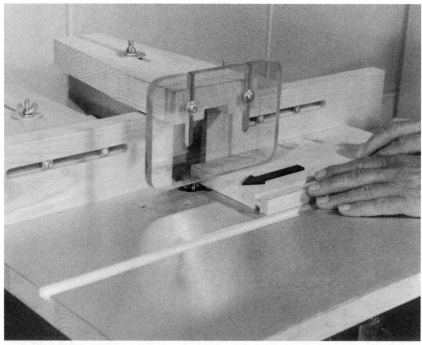

12-30 Because work must be moved against the cutter's rotation, all passes (as indicated by the arrow) must be made from right to left. Position the guard so it barely clears the surface of the workpiece.

is narrow. Accuracy and safety demand that you move such pieces with a mechanical device. Some operators use a block of wood to advance the work, but that's a secondary choice.

It's often possible to make an end-grain cut on stock that is wide enough to accommodate the number of parts that are required. For example, if you need four rails that are 2 inches wide, make the rail-shaping cut on stock that is wider than 8 inches and then saw it into individual pieces. The only thing to remember is to leave an allowance for the saw kerfs.

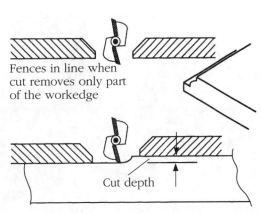

12-31 The router/shaper fences are adjusted to suit the type of cut that is being made. The fences also adjust to support the work after it has passed the cutter when the entire edge of the work is removed.

Bit rotation

Outfeed fence   Infeed fence

Feed direction
right to left

Fences in line when
cut removes only part
of the workedge

Cut depth

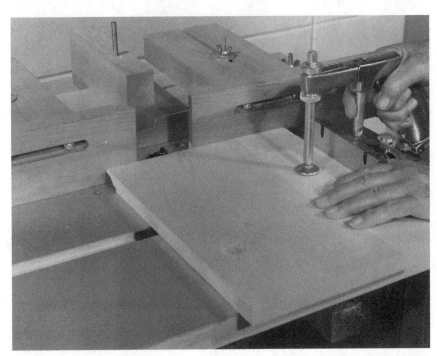

12-32 A miter gauge makes it easier and safer to make crossgrain cuts. A hold-down can be a big help. The guard, even though it is not shown here, should be used.

Figure 12-33 demonstrates another application for the technique. It would be very difficult, let alone unsafe, to end-shape small parts like the one shown in the photograph; however, if a large block was shaped first, it could be sawed into any number of identical pieces.

There is always some feathering or splintering at the end of cross-grained cuts. Therefore, when shaping two adjacent edges or all four edges of a workpiece, make the cross-grained cuts first. The final cuts that are parallel to the grain will remove the imperfections.

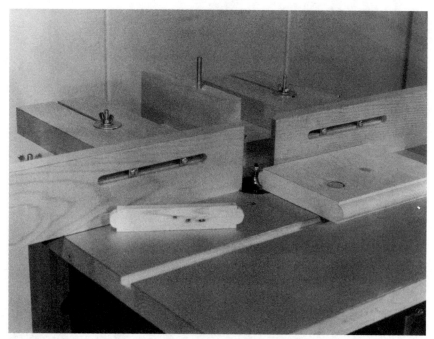

12-33 Large pieces can be shaped and then sawed into many identical parts.

FREEHAND SHAPING

Freehand should not be taken literally. It simply applies to working on circular or curved edges that can't be guided by a fence. The width of the cut and support for the workpiece during the pass are provided by the pilot, preferably a ball-bearing one, that is on the cutter (FIG. 12-34). You can see that a pilotless bit can't be used for this type of cutting. Always be sure that the arrangement provides sufficient bearing surface for the edge of the work (FIG. 12-35) and that you use the fulcrum pins.

The cut is started by bracing the work firmly against the first fulcrum pin (infeed side) and then very slowly advancing it until it contacts the cutter and receives support from the pilot (FIG. 12-36). Once the work has good support from the pilot, it may be swung free of the fulcrum pin. It's a moot point whether this is good practice or not. There is nothing wrong with continuing to use the pin for work support as long as the shape of the work

12-34 The pilot on the router bit provides bearing surface for the work and also controls the depth of the cut.

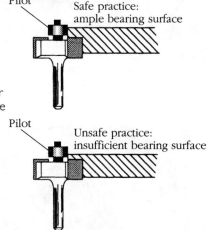

Pilot

Safe practice: ample bearing surface

12-35 Always be sure that the pilot provides enough bearing surface for the workpiece. Bits that do not have pilots cannot be used for this type of cutting.

Pilot

Unsafe practice: insufficient bearing surface

permits it. Many operators work with both fulcrum pins installed, relying on the outfeed pin to provide support at the end of the cut. The shape of the work will have some bearing on how long the infeed pin can provide support and whether both pins can be utilized.

Never try to shape pieces that are too small or too narrow for safety. When a narrow part is needed, form the shape on a board that can be hand-held safely and then saw off the part that is needed (FIG. 12-37).

IN GENERAL

Efficient cutting procedures call for moving the stock just fast enough for cutters to work freely. Forcing cuts will not speed up production. Any feed speed or cut depth that slows the router excessively, overheats the cutter, or burns the wood is obviously poor practice. All work edges must be smooth and square to adjacent surfaces initially. Always work with sharp cutters and maintain pilots and bearings in pristine condition.

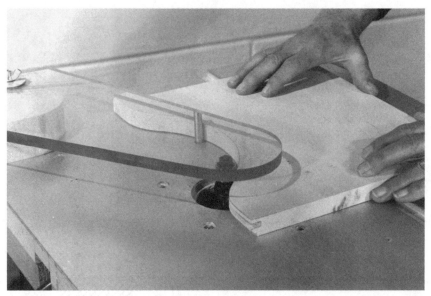

12-36 The work, after being braced against the infeed fulcrum pin, is advanced very slowly to make contact with the bearing on the cutter. Be sure to keep the work flat on the table throughout the pass and never allow your hands to come close to the cutting area.

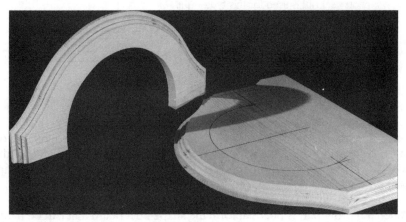

12-37 When you need narrow parts, form the shape on a large piece of stock and then saw off the area you need.

While the router supplies the power and the cutters produce necessary shapes, it's the operator's responsibility to understand procedures and to establish correct cutter-to-work relationships. This is especially important when using matched cutters like those often supplied for stile and rail assemblies (FIG. 12-38) and for other cutters like the glue joint (FIG. 12-39).

Building a library of sample cuts, like those shown in FIG. 12-40, can considerably reduce setup time on future projects. Once you have shaped

12-38 Cope and stile cutters and other bits are often offered as matched sets. You must be very careful when setting up for each of the cuts. It's often possible to use the first cut that is made as a gauge to establish the correct position for the second cutter.

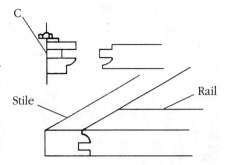

Stile and rail assembly

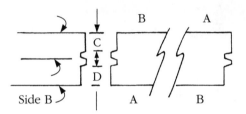

Glue joint cutter

Mark board surfaces that will be up after assembly (face side of stock)

–Cut one edge of stock with face side (A) up

–Cut edge of mating piece with face side down

NOTE: Shoulders C and D must be equal regardless of thickness of the stock

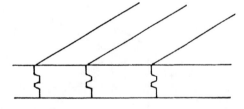

12-39 This is the procedure to use in order to get accurate mating edges when using a glue joint cutter.

a piece correctly, cut off a small section of it and save it to use as a gauge when a similar piece is required.

A lot of the cutters that must be used in a router/shaper arrangement are designed for frame and panel constructions (FIG. 12-41). While they might be used for similar results, operational procedures will differ. Some sets have matched cutters, others include individual cutters that, in a sense, are multipurpose. They can produce various elements of a shape depending on width of cut, depth of cut, and so on. It is very important then to carefully follow the instructions that are supplied with the cutters. Most manufacturers are very generous with educational materials, so study them.

12-40 Having a library of sample cuts is a big help because the pieces can be used as gauges when it is necessary to duplicate a setup.

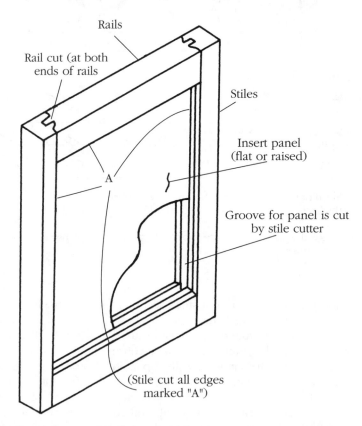

Rails

Rail cut (at both ends of rails

Stiles

Insert panel (flat or raised)

A

Groove for panel is cut by stile cutter

(Stile cut all edges marked "A")

12-41 This drawing identifies the parts of a panel assembly and tells where various cuts should occur.

13

Other ways to mount a router

The portable router can be used in a variety of ways. We've already learned that while its portability is one of its outstanding features, there are times when organizing it as an overarm machine or installing it in a shaper table is a practical way to go, but there is more. Accessories that are available from several manufacturers provide for establishing the tool on a table saw or radial arm saw.

ON A TABLE SAW

The Bosch table saw router table is designed for use on saws that have a 27-inch removable extension. Substitute the unit for the extension and you have a substantial router work surface that doesn't require additional floor space and that can utilize the saw's surface as well as its own. A predrilled, steel mounting plate with special brackets is flexible enough to accept just about any router, not just Bosch tools.

The kit includes the router table (or *leaf*) with attachment hardware, mounting plate, a wooden fence that is notched so it can be placed over a cutter when necessary, and a pivoting see-through guard. A plus feature is that the auxiliary table will accept a saber saw (portable jigsaw) as well as a router.

Inca recently introduced an auxiliary router table that is designed for mounting on Inca table saws. When the ⅜-inch-thick, clear plastic table is installed in place of the saw's extension leaf, it increases work surface by 8½ inches. The unit utilizes the saw's miter gauge, rip fence, and any Inca hold-down accessories that are normally used for sawing operations. You must drill holes for mounting the router, but that's not a bad feature since it allows working with any router you happen to own.

ON A RADIAL ARM SAW

When you combine the capability of a portable router with the flexibility of a radial arm saw, you have a concept that rivals, and in some cases surpasses, what can be done with overarm and pin-routing machines. This is so because the router is mounted in place of the saw blade, or the machine provides a means of gripping router bits so all the motions that the saw is noted for can be utilized. The router, or bit alone, depending on the arrangement, can be stationary while workpieces are moved by it, or vice versa. With workpieces clamped or held by some other means, the cutter can be raised or lowered, swung in a vertical or horizontal arc, moved in a circular path, even positioned horizontally so the bit is parallel to the tool's table. Standard router bits, with or without pilots, depending on the operation, can be used.

WAYS TO GO

Ryobi and Craftsman (Sears) have saws with auxiliary spindles that are actually router collets or threaded to receive an accessory collet. Both units will work with router bits that have a ¼-inch shank.

Inca has introduced a *router carriage* that is designed for their 810 radial arm saw. The unit, that actually replaces the motor assembly, takes little time to install. Once mounted, it can be moved to-and-fro, held in a fixed position, and go through any of the actions normally associated with radial arm sawing. It's also possible to set the router in an inverted position so that, with an additional work surface, it can function as a router table.

Special router mounting accessories are available that can be used as is or modified to suit a saw you already own. Two versions, a Craftsman and a DeWalt, are shown in FIGS. 13-1 and 13-2. Another solution is to acquire a special commercial bracket (*Router Bracket*) that consists of two semicircular cast aluminum components that embrace the router so it can be installed. This requires removal of the saw's motor from the tool's yoke. In essence, the setup transforms a sawing tool into a very flexible routing machine.

Finally, if a commercial unit can't accommodate your needs, make your own along the lines of that which is detailed in FIG. 13-3. The homemade unit is basically simple, but you can judge that it must be dimensioned and shaped exactly for the tool you will use it on. If you are not equipped for metalworking, you can make a prototype using easy-to-bend-and-drill sheet metal and have the final product made for you by a metalworking shop. It's also possible to make a sample by using stiff cardboard. The slot for the saw guard stud must be wider than necessary to rotate the unit to position the router perpendicular to the table. Use long carriage bolts to secure the router between the clamp blocks.

The saw becomes just a carrier; don't have it plugged in when using the router. Also, situate the router's cord, possibly by threading it through the yoke of the machine, where it will be safe and won't interfere with the work. Be sure the router switch is in the off position before plugging it in.

13-1 With this special router mounting accessory, only the motor of the router is used. There is enough leeway in the clamping arrangement so that almost any router can be used.

CUTTING

The direction in which you move the work must always be *against* the cutter's direction of rotation. Because the router's attitude is the same as it would be if you were hand-holding it and looking down on it, the bit's rotation is clockwise. Feeding too fast, regardless of whether you are moving the router or the work, results in inferior cuts and can harm the tool and the cutter. Conversely, being too cautious allows the bit to do more burning than cutting. Because wood densities vary, an efficient feed speed is a judgment that must be made by the operator. The usual precautions apply. If the cutter chatters or the router slows excessively, it's possible that you are cutting too deep or too fast, or that you are using a dull bit. By this time you probably have done enough work with a router to know how it should feel, sound, and move when being used efficiently. If not, some practice cutting will soon get you to the right plateau.

STRAIGHT CUTTING

Cuts that are made across the grain to produce forms like rabbets and dadoes are done with the saw that is set up in normal crosscut position (FIG. 13-4). The depth of the cut is controlled by the cutter's height above the table and the width of the cut is determined by the diameter of the bit. The

13-2 The Sears version of the router mounting accessory is also secured on the saw end of the motor, but it provides a platform for the router with its base to rest on. Routers with a 6-inch-diameter base can be used.

width of the cut is not limited by the size of the bit. You can get to any cut width with any bit simply by making repeat passes. How deep you can cut in a single pass depends on the density of the material and the horsepower of the router. A few light cuts, say between ⅛ inch and ¼ inch, will quickly guide you along the right path.

For stopped or blind cuts (FIG. 13-5), start with the bit elevated above the work and then slowly lower it, by using the tool's regular arm-height control, until it enters the work. Then move the router as you would normally. You can work to lines marked on the work, or you can control the length of the cuts by using small clamps on the tool's arm to limit how far the motor can be moved. You'll get accurate results when repeat passes are needed to deepen a cut if you start with the work clamped in position.

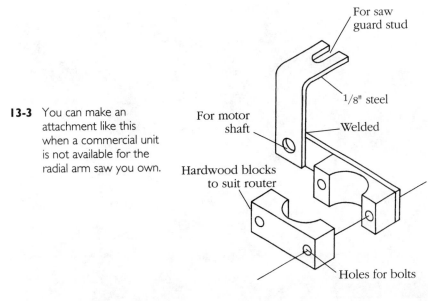

13-3 You can make an attachment like this when a commercial unit is not available for the radial arm saw you own.

For saw guard stud

1/8" steel

For motor shaft

Welded

Hardwood blocks to suit router

Holes for bolts

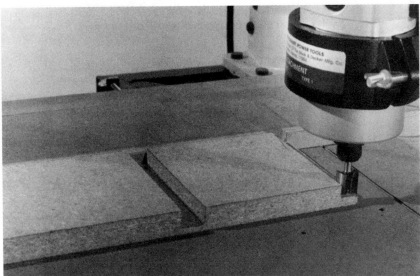

13-4 The tool is used in crosscut position for cuts like dadoes and rabbets. When the same cut is required on many pieces, use a stop block on the fence to gauge the work's position.

To do grooving or rabbeting on long edges, the router's position is secured with the tool's rip lock. The work, guided by the fence, is moved to make the cut. The setup is the same as that required when using the saw for ripping operations.

Some types of standard molding designs can be duplicated by making through or stopped cuts across stock that has been ripped to suitable width

13-5 Because the router can be raised or lowered by using the tool's arm-elevating mechanism, stopped or blind cuts pose no problem.

(FIG. 13-6). Many types of pilotless bits can be used on work like this, so results are not limited to grooves formed with a straight bit. Parts that are formed by using this particular technique can also serve as base stock that can be sawed to produce slim moldings (FIG. 13-7).

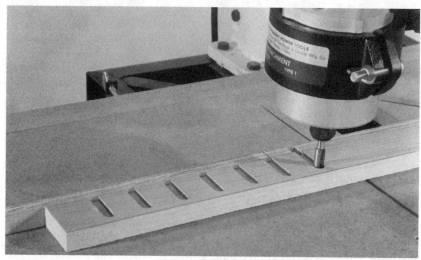

13-6 Use a typical setup to surface cut strips for molding. Results will be more dramatic if you use pilotless, decorative bits. A mark on the fence can be used to gauge spacing between cuts. Use a small clamp on the arm of the tool to limit motor travel when cuts must be stopped.

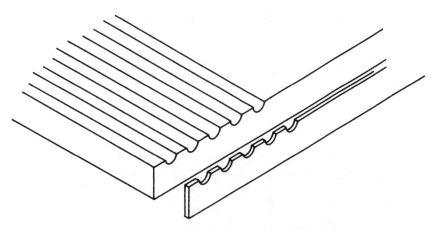

13-7 Surface cut pieces can be sawed to produce slim moldings.

Straight edges can be shaped by using a fence as a guide for the work. As shown in the example in FIG. 13-8, the router's position is locked and the work, held snugly against the fence, is moved from left to right. You can provide for the cutter's height above the table by cutting a relief area in the fence or by working with a two-piece fence. A two-piece fence is a good idea because it can be adjusted to minimize the opening around the cut area regardless of the bit that is being used.

13-8 To shape edges, lock the router in position and move the work along the fence. The work is moved from left to right. This is a good way to operate even if the bit has a pilot.

SHAPING CURVED AND CIRCULAR EDGES

Shaping curved and circular edges while hand-holding the router should be accomplished by first clamping the work, then moving the router to make the cuts. The router on a radial arm saw is stationary, so the work is moved to make the cut. The depth of the cut is controlled by the cutter's height above the table. The work is guided by an integral or ball-bearing pilot that *must* be on the bit (FIGS. 13-9 and 13-10).

13-9 Curved edges are shaped by keeping the router still and moving the work against a piloted bit. You have some control over the shape you can get from any bit because the router's height above the table is adjustable.

Circular edges can be shaped perfectly by using the pivot guidance method that is demonstrated in FIG. 13-11. The work is impaled on a nail driven up through the bottom of a plywood support that is clamped to the regular table. After adjusting the height of the cutter, hold the work firmly and bring the cutter forward to make contact. With the rip lock tightened to keep the router stationary, rotate the work—in this case in a counterclockwise direction—to make the cut.

You can use the same technique to form circular surface grooves and even to cut out discs (FIG. 13-12). A groove is just a limited depth cut. For discs, use a small diameter straight bit and, if necessary, make repeat passes until the cutter is through the work. Elevate the work on some scrap material so the bit will not cut into the saw's table.

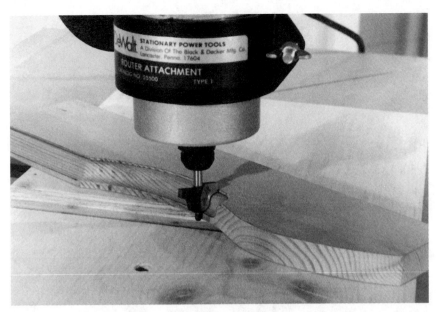

13-10 In some cases it's necessary to elevate the workpiece. A piece of plywood with a dado cut across it will serve. The dado provides extra freedom for the bit's pilot.

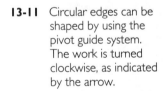

13-11 Circular edges can be shaped by using the pivot guide system. The work is turned clockwise, as indicated by the arrow.

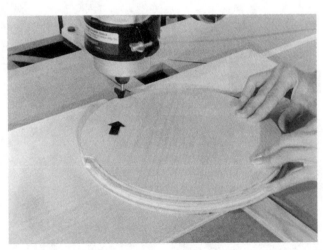

Internal routing is possible because the router is adjustable vertically by raising or lowering the tool's radial arm. This means that you can raise the router and, after placing the workpiece, lower it to position the bit for the correct cut (FIG. 13-13).

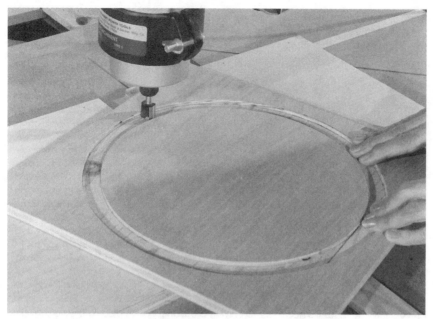

13-12 The pivot technique can also be used to form circular grooves. Discs can be formed the same way, but elevate the work to keep the bit from cutting into the tool's table.

13-13 To shape inside edges, place the work in position before lowering the router.

FREEHAND ROUTING

Freehand routing is done by manipulating the work while the router is in a fixed position (FIG. 13-14). All the precautions that were suggested for this type of operation when the router is used in routine fashion apply here. Final results depend primarily on how carefully you work. Remember that the bit tends to be contrary, following grain lines and patterns rather than your directions. Grip the work very firmly and work more slowly than you normally would. It's a good idea to outline figures first by working with a small diameter bit. The remaining waste can then be removed with a larger cutter.

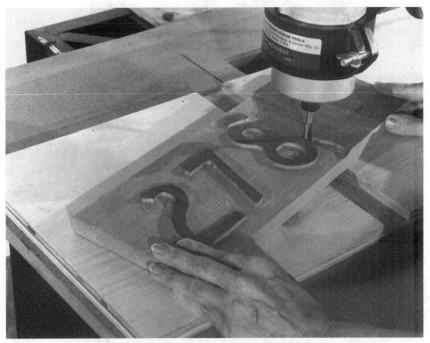

13-14 A firm grip on the work is necessary when doing freehand routing. Practice on scrap material before starting a project. Make test cuts on hard and soft woods.

HORIZONTAL ROUTING

Horizontal routing is a term that can apply to the router/radial arm arrangement because the router can be placed in the position that is demonstrated in FIG. 13-15. Because of the way the router is mounted, there is a limit to how close to the table a cutting bit can be situated. Therefore, a special table that provides an elevated work surface and that has is own fence is needed. The fence is actually a back piece that extends below the legs of the table so it can be gripped in the slot between the tool's table boards that is normally occupied by a regular fence.

Figure 13-16 is a typical example of how horizontal routing can be utilized. The work is held or clamped to the table, and the bit does the cutting

13-15 When the router is set in horizontal position, the bit can't be positioned close to the tool's table. A special table that provides an elevated work surface is needed.

13-16 Forming an end groove is a typical operation made possible when the router is secured in a horizontal position.

when the router is pulled forward. Other cuts that can be made by using this setup include rabbets, tenons, dovetail slots, and end-grain shaping. When you consider that the router can also be tilted when in a horizontal position, you can envision many other possible applications.

14

Some interesting major accessories

The portable router's versatility increases when accessories utilize the tool as a high-speed driver of a host of cutting bits. Operators can devise any number of special guides, jigs, and holders to use the tool well beyond its basic concept. This also holds true for manufacturers who make routers or unique accessories that hold a router in a particular way or supply controlled movement for the tool so it can be applied in various ways to the workpiece.

Because of the accessories, a router can be used to form various types of spirals or flutes and reeds on spindles, turn out letters and numbers on a production basis, form bowl-type projects, do bas-relief or intricate 3-D carving, and more. It isn't necessary for router owners to buy all the extras or, for that matter, even one of them. Much depends on the length to which you wish to increase the scope of your router workshop.

All the units come with very detailed instructions for efficient use and, in some cases, for correct assembly. It's crucial to study the pamphlets and to accept the information there stated. The supplier wants you to be happy with the product.

THE ROUTER CRAFTER

The Router Crafter that is shown in FIG. 14-1 is another unit in the impressive array of router accessories that are offered by Sears, Roebuck and Co. under the Craftsman brand name. As the photos indicate, the tool imitates a lathe. It has an adjustable tailstock that accommodates work of various lengths and a headstock that can be indexed. A major feature is that the headstock, and thus the work, can be rotated by means of a hand crank. Because of the way the tool can manipulate a router or the workpiece—or

14-1 The Router Crafter uses a router as a motorized lathe chisel.

both—simultaneously, you can do a variety of intriguing operations on spindles with mechanical precision. Very little assembly work is required before the accessory can be used. It can be secured directly to a bench, but I think a box-type structure, like the one shown in FIG. 14-2, is a better way to go. By adding a simple drawer, the base could provide storage for other tools, router bits, and so on.

14-2 The Router Crafter can be mounted directly on a workbench, but box-type support is more practical. It makes the unit portable and can also be used for storage.

What makes the Router Crafter so flexible is the cable drum that is situated in the headstock. Turning the hand crank causes the carriage-mounted router to move parallel to the workpiece. Small cable-mounted clamps can be positioned to limit the length of cuts. Because of the indexing mechanism in the headstock, it's easy to divide the workpiece into 2, 3, 4, 6, 8, 12, or 24 equal spaces. Spiral cuts can be made by moving the router longitudinally while the work is rotating. It's also possible to do peripheral cutting by keeping the router still while the work is turned.

After becoming acquainted with the tool, you'll find it fairly easy to do the following operations on table legs, lamp bases, posts, and other projects or project components that are not greater than 3-inch square or more than 36 inches long:

- Form straight or tapered, equally spaced flutes or beads parallel to the workpiece.
- Do left- or right-hand *roping* or *spiraling*. Diamond patterns occur when left- and right-hand cuts are combined.
- Do peripheral shaping by working with pilotless router bits.
- Form contoured spindles by organizing a setup that allows the router to follow a template that you attach to the front of the accessory.

These operations are done by following basic procedures, but you can go further by combining different types of cuts. For example, you can do peripheral cutting after a workpiece has been shaped lengthwise with flutes or reeds. Figures 14-3 through 14-6 show examples of work that can be done with the Router Crafter.

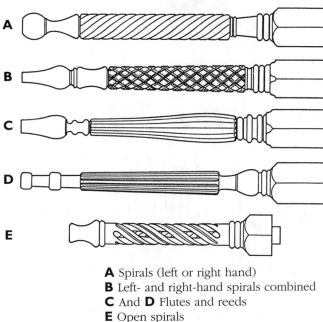

A Spirals (left or right hand)
B Left- and right-hand spirals combined
C And **D** Flutes and reeds
E Open spirals

14-3 These are examples of some of the work that can be done on the Sears Router Crafter. Diamonds appear when both left and right hand spirals are formed. Open spirals require a through concentric hole in the workpiece.

14-4 Spirals can be formed to the left or right and can be embellished by repeating a cut with a different bit. Work can be done on tapered spindles as well as straight ones.

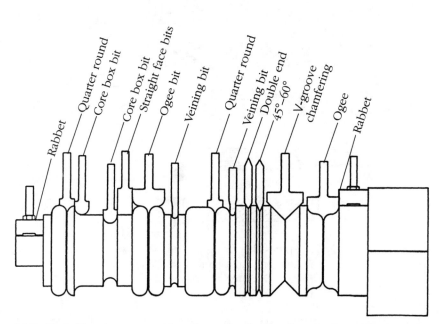

14-5 Typical beads, coves, and flats that can be produced on the Sears Router Crafter. Usually for this type of cutting, the router is held still while the workpiece is rotated.

14-6 To shape particular profiles, the router is controlled by a guide that rides against a template that is secured at the front of the machine. This is a good way to form duplicates.

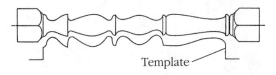

Template

THE MILL-ROUTE

The Mill-Route (FIGS. 14-7 and 14-8) produced by Progressive Technology, Inc., works something like a pantograph but has control features that help you work with ultimate precision. The accessory can be used to produce a single project or to make a template that can then be followed by the stylus to form any number of duplicate parts. It's literally true that the router, which is counter-balanced when it is mounted, will mimic any movement of the stylus, whether it is following lines of a drawing or a template. The router has a vertical lifting action that ensures that all cuts will be perpendicular to the surface of the work. A nice feature is that the router automatically lifts from the work whenever the operator removes his or her hands or from the control bar.

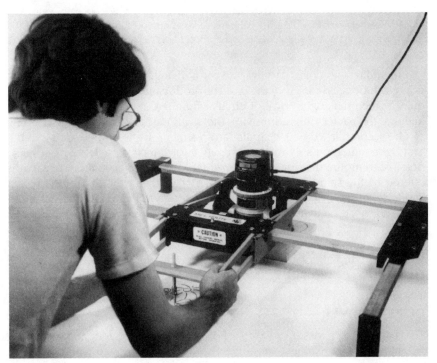

14-7 The Mill-Route is an easily controlled machine that uses a router to follow any line that is traced by the stylus.

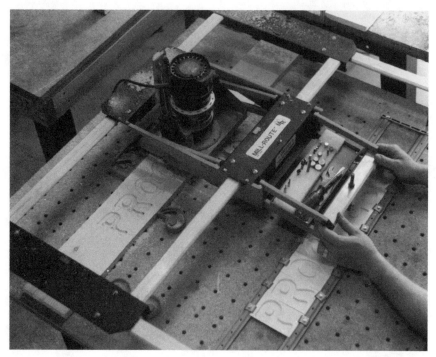

14-8 Work can be done by using ready-made templates or by designing your own. Whether free or cutting, the router is always perpendicular to the surface of the work. It lifts automatically when the control bar is released.

In addition to doing carving and forming letters and numbers, the Mill-Route can be used for grooving and for producing through cuts that are required when making an oval or circular frame. Moldings can be shaped by keeping the router still and moving the work past the cutter. Edge shaping can be accomplished with a similar arrangement. It's recommended that the tool be mounted on a 44-inch-square sheet of ¾-inch particleboard. It's also possible to mount the unit on a special table, an extra cost accessory that is offered by the same manufacturer (FIG. 14-9). Either way, the tool covers a 20-inch × 24-inch area, which is capacity enough to make large signs and to work on most cabinet and door panels. The universal mounting plate is guaranteed to fit your router.

The company has started to offer ready-made templates. Some letter sets and carving templates that can, for example, be used to decorate cabinet doors, picture frames, or clock faces are currently available.

DUPLI-CARVER

The Dupli-Carver, produced by Laskowski Enterprises, Inc. and shown in FIG. 14-10, is more than just a router accessory. It is equipped with a standard ⅝-horsepower router and a speed control and should be thought of as a complete unit. The tool is sophisticated, but can be quickly mastered by anyone.

14-9 You can mount the Mill-Route on a panel of ¾-inch particleboard or on a work table that is offered at extra cost. It's also possible to make your own separate stand.

14-10 The Dupli-Carver is supplied with its own ⅜-horsepower router. Intricate carvings in-the-round are possible when the blank and model are mounted on synchronized turntables.

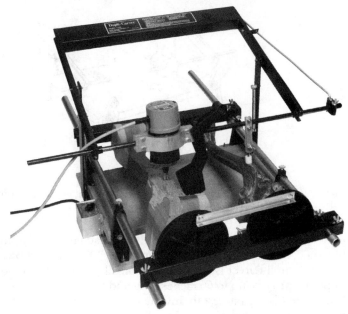

The "secret" of the machine is the way the router can be manipulated to duplicate any configuration that is traced with the stylus. As demonstrated in FIG. 14-11, the router floats through various actions almost as if it were an extension of your hand. This, plus being able to mount workpieces on turnta-

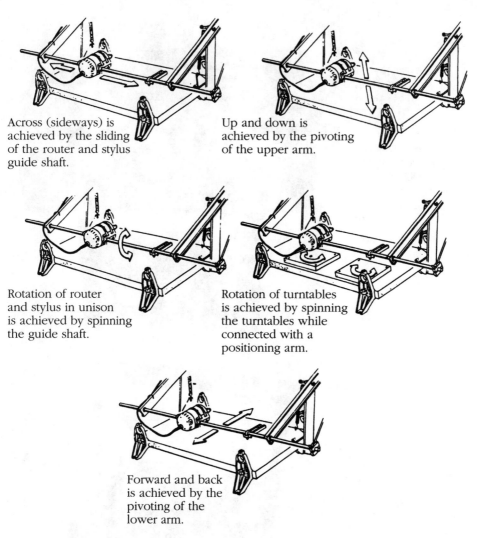

Across (sideways) is achieved by the sliding of the router and stylus guide shaft.

Up and down is achieved by the pivoting of the upper arm.

Rotation of router and stylus in unison is achieved by spinning the guide shaft.

Rotation of turntables is achieved by spinning the turntables while connected with a positioning arm.

Forward and back is achieved by the pivoting of the lower arm.

14-11 The various actions of the Dupli-Carver are what enable the router bit to follow intricate undercuts and contours.

bles—two of which are supplied with the machine, is what makes it possible to do, step-by-step, the kind of intricate 3-D carving shown in FIG. 14-12.

The Dupli-Carver is not limited to small projects. It can be organized to duplicate projects up to 66 inches long and 8 inches in diameter (FIG. 14-13). What might be of interest to anyone doing production work is that several machines can be organized in tandem so that a single operation can produce two, three, or four objects at a time and they can be as large as 32 inches in diameter and 40 inches tall. A special line of router bits is offered, many of them designed for duplication of very fine detail (FIG. 14-14).

Other accessories include a variety of lettering templates, various sizes of styli, sanding cones, a special unit for sign making, and more.

14-12 This bust of Lincoln demonstrates the various stages of roughing and adding detail to a piece of work. Results are excellent when you take your time and work carefully.

14-13 The Dupli-Carver can handle projects up to 66 inches long and 8 inches in diameter. Among the commercial people who use the tool are gunsmiths, antique restorers, and makers of musical instruments.

ROUTER-RECREATOR

With the Router-Recreator pictured in FIG. 14-15, you can carve 3-D figures up to about 8 inches tall (FIG. 14-16) and also make signs, do fluting on spindles, shape edges, form uniform depth or tapered grooves, do bas-relief carving, and even do some panel decorating. It's a lot to expect from a single router accessory, but the tool doesn't let you down.

Any router that has a 3-inch to 3¾-inch diameter motor body can be mounted on the counter-balanced shaft that also holds the stylus. The shaft rides on pulleys that are controlled with steel cables. The router can be

Part # Size	Function

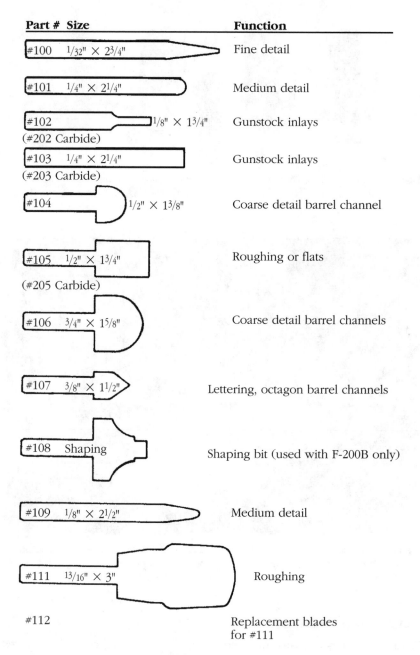

#100 1/32" × 23/4"	Fine detail
#101 1/4" × 21/4"	Medium detail
#102 1/8" × 13/4" (#202 Carbide)	Gunstock inlays
#103 1/4" × 21/4" (#203 Carbide)	Gunstock inlays
#104 1/2" × 13/8"	Coarse detail barrel channel
#105 1/2" × 13/4" (#205 Carbide)	Roughing or flats
#106 3/4" × 15/8"	Coarse detail barrel channels
#107 3/8" × 11/2"	Lettering, octagon barrel channels
#108 Shaping	Shaping bit (used with F-200B only)
#109 1/8" × 21/2"	Medium detail
#111 13/16" × 3"	Roughing
#112	Replacement blades for #111

14-14 Pictured are the various types of cutting bits that are used with the Dupli-Carver. Part numbers are those of the manufacturer. All are high-speed steel.

moved backwards or forwards, from side to side, and can be tilted. The actions are controlled individually or can be combined. This allows the router bit to imitate how the stylus moves so it can follow intricate configurations.

14-15 The Router-Recreator works with a router and stylus that mount on the same counter-balanced shaft. The shaft rides on wheels that are controlled with steel cables.

14-16 This plaque was made by using a chess piece as a model.

The weight that supplies the counter-balancing is adjustable, making it easy to provide a light touch regardless of the router's weight.

The machine can be used to do 3-D carving as demonstrated in FIG. 14-17. Both the model and the blank are secured on mounting blocks that can be rotated 360 degrees. They must be turned in unison. This is accomplished by following the corresponding marks on the mounting blocks and the tool.

14-17 Model and blank are mounted on blocks that can be rotated 360 degrees. It's important for the blocks to be turned in unison.

Many types of conventional router bits can be used with the Router-Recreator, but for carving operations, it's wise to buy the special set shown in FIG. 14-18. The set comes with the machine or can be purchased separately on its own. Buying the products together is a good idea because the price is less than if the tools were purchased individually. Also available for the machine is a foot-activated switch that lets you keep both hands free to control the router. Figures 14-19 through 14-24 show some ways in which this interesting router accessory can be used.

THE TRIMTRAMP

The Trimtramp, shown doing a dadoing operation in FIG. 14-25, is a Canadian product that is widely available in the United States. The product has an unusual name, but it's a pretty fair accessory for a portable router. The product was designed for a portable circular saw so the basic unit must be on hand before the recently introduced, special router kit can be utilized.

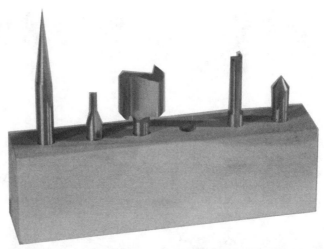

14-18 Many types of router bits can be used with the Router-Recreator, but for carving purposes, it's essential to have the cutters in this set.

14-19 Some panel decorating is possible with the Router-Recreator. The stylus, indicated by the arrow, follows the outline of a template; the router duplicates the moves.

The kit consists of a mounting plate with adjustable clamps that will accommodate any router, a fence, movable hold-down springs, a stop, and all necessary hardware. If you own a Trimtramp and add the router kit, you don't need anything else except, of course, a router and bits.

14-20 To create a raised panel area, you can switch to another stylus tip and a larger diameter bit.

14-21 Straight fluting can be done by moving the stylus along kerfs that are cut into a piece of wood. You can follow a marked line, but the kerfs provide more control. For tapered cuts, place a height block under one end of the workpiece.

14-22 Sign making is another possibility with the Router-Recreator. You can guide the stylus with stencils or with templates.

14-23 Some amount of edge shaping can be done. A piloted bit is used while the stylus acts as a height control. Always be sure that workpieces are firmly clamped.

14-24 Fluting can also be done on spindles. The stylus acts only as a height control. Note the C-clamp being used to secure the work.

14-25 A new kit for the Trimtramp allows routers to be used in place of a circular saw. The mounting plate slides smoothly between sturdy aluminum bars. Here, the work is still while the router is moved to form dadoes. Note the adjustable spring hold-downs.

314 Some interesting major accessories

Basically, the unit is used two ways—the work is held securely while the router is moved or the router is kept still while the work is moved (see FIGS. 14-26 and 14-27).

14-26 A stop (arrow) controls the travel of the router for stopped cuts. Operations like this are more convenient to do when a plunge router is used. An auxiliary particleboard table about ⅝ inch × 16 inches × 30 inches is supplied by the user.

Another portable circular saw product that can be used for router operations is shown in FIG. 14-28. In this case, the mounting plate that is supplied serves either a saw or a router.

CRAFTSMAN MILLWORKS

The Millworks, shown in use in FIG. 14-29, can be used to duplicate existing molding forms or to create original ones. The product has a mounting plate that will accept routers with up to a 6-inch-diameter base. Work capacity ranges from 1½ inches to 6 inches wide and from ½ inch to 2 inches thick. There is no limit on work-length.

The unit has a depth-stop scale and is adjustable so the carriage that holds the router can be preset for a particular travel distance. The router can be moved directly into the work to produce dentil-type cuts, moved longitudinally to create fluting or grooving (FIG. 14-30), or used to produce more complex molding forms on stock edges.

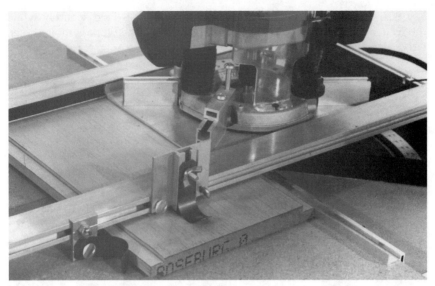

14-27 Longitudinal cuts are made by feeding the work while the router is locked in position. Cuts can be made "in the field" as well as on edges. It's a good idea to buy an extra set of hold-downs so they can be used on each of the aluminum bars.

14-28 Routers can be used on the Sawtrax, another product that was designed for portable circular saw use. In this case, the supplied mounting plate also accommodates a router.

14-29 The Millworks, designed for reproducing standard moldings or for creating original ones, will accept most routers with a 6-inch base. The carriage that grips the router can be moved forward a controlled distance for dentil-type cuts, or the entire unit can be moved along stock edges for longitudinal cuts.

14-30 A brief view of some of the work that can be done with the Millworks.

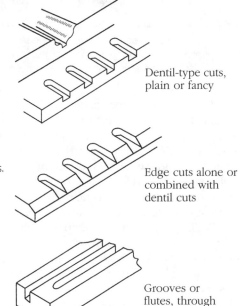

Dentil-type cuts, plain or fancy

Edge cuts alone or combined with dentil cuts

Grooves or flutes, through or stopped

CRAFTSMAN ROSETTE MAKER

The Rosette Maker, shown mounted on a Craftsman router/shaper table in FIG. 14-31, provides a fairly simple way to produce decorative pieces like those in FIG. 14-32. The product is supplied with parts for mounting on Craftsman tables, but it's reasonable to suggest that with some modification of the attachment method, it might be suitable for other tables.

14-31 The Rosette Maker fits over the router bit. The workpiece, mounted on a faceplate, is secured to the end of the indexing shaft and is rotated by the crank. The product is made for Craftsman router/shaper tables, but with some attachment adjustments should be usable on other tables.

The Rosette Maker concept is fairly straightforward. The workpiece is mounted on a faceplate as it would be for routine lathe work (FIG. 14-33). Then the faceplate is secured to the end of the index shaft that projects inside the tool's housing. After the unit is secured to the table, the router is turned on, and the work is rotated by turning the handle on the index shaft. Adjustments are provided for controlling the depth of the cut and its radius.

It doesn't take long to master the procedure, but follow the practice of making some test pieces before you produce components for projects.

Well, again, it's time to stop. One of the problems with router shop talk is that it can go on forever. New ideas and new applications are daily events. By the time this book is in print there will probably be more accessories, more bits, and more router innovations. The information included in this second edition, though, should have you prepared for whatever comes along. Until next time then—safe and happy woodworking.

14-32 Example cuts produced with the Rosette Maker.

14-33 Workpieces, square or round, are mounted concentrically on the face plate with the ⅝-inch-long screws that are provided. If workpieces are less than ¾ inch thick, it's necessary to use a spacer between the work and the faceplate. The total thickness should be from ¾ inch to 1 inch.

Suppliers

American Woodcraft Tools, Inc.
10420 Kinsman Road
Newbury, OH 44065-0200
216-564-9600

Byrom router bits, miniature bits

Black & Decker
10 N. Park Drive
Hunt Valley, MD 21030
1-800-762-6672

Routers, bits, accessories

Bosch Power Tool Corporation
100 Bosch Boulevard
New Bern, NC 28561-2217
1-800-334-4151

Routers and trimmers, complete line of router accessories

CMT Tools
5425 Beaumont Center Boulevard
Tampa, FL 33634
813-886-1819

Extensive bit assortment, router tables and other accessories

Dremel
4915 21st Street
Racine, WI 53406
414-554-1390

Compact power tools with accessories and bits

Freud USA, Inc.
218 Feld Avenue
High Point, NC 27264
919-434-3171

Extensive bit assortment, heavy-duty routers, drilling tools

Garret Wade Co.
161 Avenue Of The Americas
New York, NY 10013

Inca products, general woodworking tools and supplies

Hitachi Power Tools USA, Ltd.
4487 E. Park Drive
Norcross, GA 30093
1-800-548-8259

Routers and basic accessories, bits

Keller & Co.
1327 I Street
Petaluma, CA 94952
707-763-9336

Keller dovetail system

Laskowski Enterprises Inc.
4004 W. 10th Street
Indianapolis, IN 46222
317-243-7565

Dupli-Carver

Leichtung Workshops
4944 Commerce Parkway
Cleveland, OH 44128
1-800-321-6840

Router table, bits, accessories

Leigh Industries Ltd.
P.O. Box 357
Port Coquitlam, B.C.
Canada V3C 4K6
604-464-2700

Dovetail jigs, special bits, universal router base kits

Makita USA, Inc.
14930 N. Northam Street
La Mirada, CA 90638
714-522-8088

Routers and trimmers, bits, basic router accessories

Milwaukee Electric Tool
13135 W. Lisbon Road
Brookfield, WI 53005
414-783-8311

Routers, bits, accessories

MLCS Ltd.
P.O. Box 4053
Rydal, PA 19046
1-800-533-9298

Router speed control, foot switch

Nordic Saw & Tool Manufacturers
2114 Divanian Drive
Turlock, CA 95381-1128

Large assortment of conventional and special router bits

Oak Park Enterprises Ltd.
Box 280
Elie, Manitoba R0H 0H0, Canada
1-204-353-2692

Router accessories and bits, router video series

Porta-Cable Corporation
4825 Highway 45 North
Jackson, TN 38302-2468
901-668-8600

Routers and trimmers, bits, dovetail templates and machines, mortise tenon jig

Porta-Nails Inc.
P.O. Box 1257
Wilmington, NC 28402
919-762-6334

Router tables, panel templates, overarm accessory

Progressive Technology, Inc.
P.O. Box 98
Stafford, TX 77477-0098
713-721-3351

Mill-Route

RBI Industries, Inc.
1801 Vine Street
Harrisonville, MO 64701
1-800-487-2623

Router tables and accessories

Router Bracket Co.
P.O. Box 533
Richmond, VA 23204

Router mount for radial arm saw

Safranek Enterprises, Inc.
4005 El Camino Real
Atacadero, CA 94322
805-466-1563

Panel routing products, conventional and special router bits

Sears, Roebuck and Co.
Sears Tower
Chicago, IL 60684
Tool catalog: 1-800-366-3000

Routers, tables, bits, basic and novel accessories

Skil Corporation
4300 W. Peterson Avenue
Chicago, IL 60646-5999
312-286-7330

Routers and trimmers, bits, router table and accessories

Tinkerdell Inc.
P.O. Box 1170
Kennesaw, GA 30144
404-426-4946

SawTrax saw and router accessory

Trimtramp Ltd.
151 Carlingview Drive
Etobicoke, Ontario
Canada M9W 5S4
416-798-3160

Router kit for Trimtramp circular saw jig

Woodhaven
5323 W. Kimberly Road
Davenport, IA 52806
319-391-2386

Complete router accessories

Index

bits, *cont.*
  groove bit, **40**
  key slot, 39
  ogee, 39
  round edge, 39
  securing, 67
  shaper, **264**
  slot cutters, **40**
  slotting, 39
  straight, 39
  V-groove, 39
  veining, 39
Black & Decker, 2, 9
blind and through cuts, 95-97
blind cut, 85, **97**

C

cabinet door lip bit, **40**
Carter, R. L., 1
carving, 219-222
  jig for, 220
  motor holding stand, construction,
    221
  preshaping, **222, 223**
  sculptures, **222**
centering guides, joints for, 133-135
chain-making techniques, 249-251
  steps for, 249-251
chair seat, 214-215
  hollowing, 214, **215**
  template, **215**
chamfer bit, 39, **38**
chess table, inlay, **234**
chuck, split-collect, 6
circles and curves, 99-105
circular and curved edges, mounted
    cutting, 294-297
circumferential cuts, motorized lathe
    chisel, **259**
clamp guide, **90, 91, 92**, 90-92
clamp strip, 78, 237
classic pilotless bit, **45**
collets, 20, 64-66, **66**
combination bit, **40**
commercial jig, joints with, 139-141
cope cutter, shaper bit, **264**
core box pilotless bit, **45**
corner biscuit joint, **167**
corner chisel, 238
cove and bead bit, 39
cove bit, 37-39, **38**
cove-type cuts, **260**
Craftsman
  2-horsepower, **14**
  Bis-Kit accessory, 166

  sizes, **166**
  Millworks, 315-317, **316, 317**
  Rosette Maker, 318-139, **318, 319**
  router/shaper table, **271**
    starting pin, 271
    use, **272**
cross lap joint, **121**
curved and circular edges, 294-297
  mounted cutting, 294-297
curves and circles, 99-105
cuts, 65, 85-98
  blind, 85
  depth and width of, 65
  stopped, 85
  straight, 85-98
  through, 85
cutter mounts bit, **40**
cutters, carbide bits, 41, **41**
cutting, 289
cylinder-type work, **256**

D

"D" handles, **10**
dado cut, 95, **97**
dado joint, **121** , 124-127
dado-rabbet joint, **121**
dado-rabbet lock joint, **121**
dadoes, with dovetail, 142-144
depth and width of cut, 65
depth gauge, **68**
design, bit/, 42-44
door lip, shaper bit, **264**
door shop bits, use, **266**
doors, hanging, tips for, 238
double bearing flute bit, **40**
double-end vee pilotless bit, **45**
double-faced tape, 79
dovetail bit, 39
dovetail dado joint, **121**
dovetail jigs, 150-165
  leigh, 150
  other types, 150-165
  ready-made, 144-150
  woodmachine company, 155-160
dovetail joint, 141-142, 163-165
  expert, 141-142
  joint master system, 163-165
dovetail lap joint, **121**
dovetail pin, overarm pin router, 195
dovetail templates, 160-163
  Keller, 160-163
  use of, 162
dovetail, overarm pin router, 195, **195**
dowel tenoning jig, 199-200, **200**
Dremel Moto-Tool, 21-24, **22**

ABOUT THE AUTHOR

R.J. De Cristoforo has written more than 30 books about woodworking and other related how-to topics. Called the "Dean of home workshop writers" by *Mechanix Illustrated*, he is a frequent contributor to such magazines as *Workbench Magazine, Popular Science,* and *Popular Mechanics.* De Cristoforo's articles have been featured in *Family Handyman, Better Homes & Gardens, Fine Woodworking,* and *Popular Woodworking,* and he has served as a consultant to Stanley, Sears-Roebuck, and Black & Decker, among other companies.

Although the workshop is his milieu, De Cristoforo has had a writing career for more than 50 years, publishing books and articles not only on woodworking topics but also in fiction and poetry. His books for McGraw-Hill include: *Woodworking for Beginners, The Band Saw Book, Gifts from the Workshop,* and *Jigs, Fixtures and Shop Accessories.*

Other Bestsellers of Related Interest

ALL THUMBS GUIDE TO PAINTING, WALLPAPERING, AND STENCILING
—*Robert W. Wood*

Make your home look beautiful with a fresh coat of paint or new wallpaper. Wood shows you how to add a decorative touch to any or all rooms in the house. You'll learn how to patch holes in drywall, choose the right paint or wallpaper, apply the paper and clean up, use stencils for dramatic effect, and choose and use the tools of the trade. This helpful guide gives you easy-to-follow, step-by-step instructions, clear, how-to line drawings *for each step*. . . a convenient lay-flat binding . . . detachable tip cards with safety precautions, troubleshooting steps, and shopping lists . . . and an all-inclusive glossary of terms. 144 pages, 180 illustrations. Paperback with lay-flat binding. **Book No. 4060, $9.95 paperback only**

HOME PLUMBING ILLUSTRATED
—*R. Dodge Woodson*

Save money, time, and frustration by doing your own home plumbing. This in-depth book gives you the confidence to tackle almost any residential plumbing job imaginable. From the absolute basics to advanced plumbing techniques, it covers: tools and materials; reading blueprints and designs; plumbing layouts and material takeoffs; cost estimating and material acquisitions; underground plumbing; the drain waste-and-vent system (DWV); installing fixtures, water pumps, and conditioners; and passing an inspection. 288 pages, 200 illustrations. **Book No. 4163, $14.95 paperback only**

THE COMPLETE BOOK OF HOME INSPECTION—2nd Edition—*Norman Becker, P.E.*

Evaluate a property inside and out. To find problems when inspecting a new home or when maintaining your present home, consult this valuable guide for advice that's guaranteed to take the guesswork and stress out of home inspection for the buyer *and* the owner. Now updated to cover current building materials, construction techniques, and home heating, electrical, and plumbing systems, it walks you through every square inch of a house and shows you how to determine its soundness. 288 pages, 155 illustrations. **Book No. 4100, $12.95 paperback only**

ALL THUMBS GUIDE TO HOME ENERGY SAVINGS—*Robert W. Wood*

Slash your energy consumption in half with only a few of the simple, inexpensive home improvements discussed in this hey-I-really-can-do-this book. You'll discover how to seal doors and windows; insulate attics, basements, and crawl spaces, insulate water heaters and pipes; stop faucet and toilet leaks; install energy-efficient lighting, attic fans, and roof turbines; use central heating and cooling more efficiently; and much more. 144 pages, at least 150 illustrations—two color throughout. Paperback with lay-flat binding. **Book No. 4244, $9.95 paperback only**

Look for These and Other TAB Books at Your Local Bookstore

To Order Call Toll Free 1-800-822-8158
(24-hour telephone service available.)

or write to TAB Books, Blue Ridge Summit, PA 17294-0840.

Title	Product No.	Quantity	Price

☐ Check or money order made payable to TAB Books

Charge my ☐ VISA ☐ MasterCard ☐ American Express

Acct. No. _____ Exp. _____

Signature: _____

Name: _____

Address: _____

City: _____

State: _____ Zip: _____

Subtotal	$ _____
Postage and Handling ($3.00 in U.S., $5.00 outside U.S.)	$ _____
Add applicable state and local sales tax	$ _____
TOTAL	$ _____

TAB Books catalog free with purchase; otherwise send $1.00 in check or money order and receive $1.00 credit on your next purchase.

Orders outside U.S. must pay with international money in U.S. dollars drawn on a U.S. bank.

TAB Guarantee: If for any reason you are not satisfied with the book(s) you order, simply return it (them) within 15 days and receive a full refund.

BC